Managing Interdisciplinary Projects

T0174154

Construction, architecture and engineering projects are complex under-takings, involving a temporary grouping of people and companies, with different agendas and experience, coming together to achieve a project goal. This book investigates the dynamics of the relationships between individuals involved in architecture, engineering and construction projects. It combines a structured theoretical framework, derived from social psychology and mainstream management theory, with case studies and research from the built environment sector. Focusing on how people interact, communicate and work together, it examines how best to manage the interdisciplinary relationships that form and reform during the project life cycle.

The book covers vital areas of project management, whose importance has recently come to be recognised, and will be valuable for students at both undergraduate and graduate level. Practitioners will also find it a useful insight into the social aspect of project management, with implications and applications that apply to all projects in the built environment sector.

Stephen Emmitt is Professor of Architectural Technology in the Department of Civil and Building Engineering at Loughborough University, UK. He formerly held the Hoffmann Chair of Innovation and Management at the Technical University of Denmark.

Managing Interdisciplinary Projects

A primer for architecture, engineering and construction

Stephen Emmitt

 Routledge
Taylor & Francis Group

LONDON AND NEW YORK

First published 2010
by Spon Press

Published 2015
by Routledge
2 Park Square, Milton Park, Abingdon, Oxon OX14 4RN

Simultaneously published in the USA and Canada
by Routledge
711 Third Avenue, New York, NY 10017

Routledge is an imprint of the Taylor & Francis Group, an informa business

Typeset in Sabon by Prepress Projects Ltd, Perth, UK

This publication presents material of a broad scope and applicability.
Despite stringent efforts by all concerned in the publishing process,
some typographical or editorial errors may occur, and readers are
encouraged to bring these to our attention where they represent
errors of substance. The publisher and author disclaim any liability,
in whole or in part, arising from information contained in this
publication. The reader is urged to consult with an appropriate
licensed professional prior to taking any action or making any
interpretation that is within the realm of a licensed professional
practice.

British Library Cataloguing in Publication Data
A catalogue record for this book is available from the British Library

Library of Congress Cataloging in Publication Data
Emmitt, Stephen.
  Managing interdisciplinary projects : a primer for architecture,
engineering, and construction / Stephen Emmitt.
  p. cm.
  Includes bibliographical references and index.
  1. Project management. 2. Architecture. 3. Engineering. 4. Building. I. Title.
  T56.8.E46 2010
  658.4'04—dc22
  2009044953

ISBN10: 0–415–48170–8 (hbk)
ISBN10: 0–415–48171–6 (pbk)
ISBN10: 0–203–88533–3 (ebk)

ISBN13: 978–0–415–48170–0(hbk)
ISBN13: 978–0–415–48171–7(pbk)
ISBN13: 978–0–203–88533–8(ebk)

Contents

Introduction   1

1 Interfaces   7
   Interdisciplinary 8
   Fundamentals 9
   Interdisciplinary project teams and groups 16
   Fundamentals of group interaction 21
   Recognising and managing interfaces 23
   Practical challenges 25
   End of chapter exercises 26

2 Communication   27
   Fundamentals 28
   Interdisciplinary project communication 31
   Language 36
   Media 38
   Electric theatre 40
   Practical challenges 42
   End of chapter exercises 44

3 Trust   45
   Fundamentals 47
   Initial trust 49
   Trust within temporary project organisations 50
   Corruption and ethics 51
   Values 54
   Value 56
   Practical challenges 59
   End of chapter exercises 60

4 Discussions 61

Fundamentals 62
Meetings and workshops 64
Meetings 66
Making meetings work 68
Facilitated workshops 72
Making workshops work 73
Combined meetings and workshops 74
Practical challenges 75
End of chapter exercises 75

5 Decisions 77

Fundamentals 78
Group decision making 81
Brainstorming and creative approaches 85
Recognising and avoiding groupthink 87
Design changes 88
Practical challenges 89
End of chapter exercises 90

6 Context 91

Fundamentals 92
Client context 93
Physical context 97
Social context 99
Leadership context 101
Process management context 104
Practical challenges 106
End of chapter exercises 106

7 Assembly 109

Fundamentals 110
Procurement choices 112
Selection of the most appropriate participants 115
Building effective relationships 119
Practical challenges 122
End of chapter exercises 123

8 Development 125

Fundamentals 126
Group development 127
Inconsistent membership 130

Conflict 132
Occupational stress and burnout 137
Project closure 138
Practical challenges 139
End of chapter exercises 141

9 Learning   143
Fundamentals 144
Reflection in action 148
Learning from projects 149
Learning from products 152
Evidence-based learning 154
Action learning and action research 156
Storytelling 158
Practical challenges 159
End of chapter exercises 160

10 Implementation   161
Context 162
Underlying philosophy 162
The values-based process model 164
Interaction via workshops 166
Leadership – the process facilitator's role 169
Application 170
Discussion 172
Reflection 173
Conclusion 174
End of chapter exercises 175

References   177
Index   187

Introduction

A project is a vehicle to bring about change. In construction that change is usually a material change, for example a new building, bridge or road, or some form of alteration, refurbishment or repair of an existing structure. Projects can also bring about non-material changes, for example a change to the way in which we do our work. Whatever their overall goal, projects are temporary events lasting for a few months or for many years. They have often a vaguely defined start and, by comparison, a clearly defined finish. Their most distinctive feature is that each is unique, shaped by the context in which it is set and by the diversely skilled individuals and organisations contributing to it.

To realise any project, be it large or small, complex or simple, requires the interaction of a variety of organisations and individuals with complementary knowledge, skills and abilities. It is the manner in which these project contributors interface during the life cycle of the project that determines how successful or otherwise the project will be. Being mainly task-based, the commonly held team goal is to complete the project on time, within budget and to pre-determined quality standards. This usually requires the application of a considerable amount of technical know-how to resolve problems and hence achieve a successful outcome. However, technical skills alone are not sufficient; the contributors must also possess the necessary social skills to be able to work together efficiently and effectively. Social interaction, be it face-to-face or at a distance by means of information communication technologies (ICTs), is an important facet of successful projects.

The uniqueness and uncertainty that comes with each new project can be highly stimulating, challenging, exciting and rewarding. The task has to be defined, the participants have to be selected and new relationships have to be established, and in some cases re-established. The one-off nature of projects is a cause for celebration, an opportunity to engage in creative thinking, forge new working relationships and enhance one's knowledge. This occurs in a dynamic project environment that should encourage innovation and collaborative working between complementary disciplines. This requires a certain amount of creative management, strong leadership from the project manager and commitment from all participants. Unfortunately not all projects reach their objectives and it is common to find their bespoke nature cited as an excuse for poor performance; which is a very poor excuse given that uniqueness is common to all projects. Failure to deliver objectives to agreed standards, on time and within budget, is usually a result of poorly conceived projects in which the participants find it difficult to interact effectively with some or all of their fellow contributors. Often the root cause is the incompatibility of the participants, organisations and/or systems, compounded by indecisive leadership and poor communication, not the uniqueness of the project *per se*.

The project manager's responsibility is to manage the project participants with the sole objective of completing the project to negotiated, agreed and clearly specified targets (quality and cost) within finite parameters (time and resources). To do this successfully it is necessary to assemble and direct people from diverse and complementary backgrounds using the most appropriate managerial framework, tools and communication media. Management is indirect in multidisciplinary projects. Participants work for, and are remunerated by, their parent organisations, hence the project manager has little direct control over the actions of individuals. He or she needs to demonstrate decisive leadership skills and be able to motivate individuals to contribute in an effective and timely manner. In essence the project manager is concerned with managing the metaphorical space between organisations and individuals, i.e. he or she is concerned with managing boundaries and interfaces. Interdependency and associated uncertainty of relationships and position within projects help to contribute to a dynamic, challenging and exciting environment. It also means that some effort is required in trying to map and then manage relationships to add value to all stakeholders. The effectiveness of this temporary social system is highly dependent on the ability of individuals and groups to work harmoniously, using their collective skills and knowledge to deliver a product that would be impossible otherwise. To borrow a dictum from architecture – the whole should be greater than the sum of the parts.

Whatever the approach taken to the management of interdisciplinary projects it is necessary to address why, when and how individuals and organisations come together, from which it is then possible to implement appropriate processes. The effectiveness of the interactions will affect the success of the project and will also influence the productivity and profitability of the organisations participating. It is the transactions between the project participants that add value to the projects and to their organisations. In the management field there is a growing recognition of the importance of people and a move away from production line thinking to more creative, dynamic and responsive approaches. The small but growing body of literature on creative management recognises the importance of creativity, culture, values and emotional intelligence (EQ). Creative management is less concerned with systems and procedures, more with individuals and their ability to apply their knowledge, skills and competences efficiently. Successful project managers know how to work with people and systems; they understand the importance of assembling the right people for the required work and using the most appropriate systems for a given context. They also appreciate the importance of getting things in place before the work starts and providing appropriate leadership throughout the various stages of the project.

This book is a primer, a collection of ideas for thinking about how to manage interdisciplinary projects. The focus is on ideas that can be used to understand the complexities of temporary project environments. The contents provide insights into how individuals interact within temporary interdisciplinary project organisations to realise the project goal(s). The aim is to introduce the main theoretical and practical factors affecting the successful management of interdisciplinary construction projects. The objective is to help students and professionals be better prepared for the complex and exciting world of architecture, engineering and construction (AEC) by understanding the social aspects of projects. Attention is on how people interact, communicate and work together; and how best to manage the interdisciplinary relationships that form (and reform) during the project life cycle. Emphasis is on developing a deeper understanding of how the success of AEC projects is determined by the collective actions of temporal groupings of individuals and organisations.

Chapters are presented in a logical and progressive sequence. Fundamental and generic factors common to interdisciplinary AEC projects are addressed in the first five chapters. This sets the scene for dealing more specifically with the management of single projects, which is covered in Chapters 6–9. The final chapter illustrates the practical application of some of the issues discussed in the book by way of a substantial case study. End of chapter exercises are included to aid independent learning. These are designed to be considered individually and/or discussed in small groups as part of a tutorial exercise.

- Chapter 1 – *Interfaces*. Understanding how individuals, groups and teams interface and work collectively is crucial to appreciating the dynamics of temporary interdisciplinary project groupings. Chapter 1 starts by addressing a number of fundamental tenants of temporary project organisations (TPOs) and interdisciplinary working. This involves the fundamentals of group interaction, the ability to recognise boundaries and the management of interfaces.
- Chapter 2 – *Communication*. Interdisciplinary management relies on interpersonal communication skills to build and maintain effective and efficient working relationships. The ability to put people together who can communicate effectively and make informed decisions within an appropriate communications infrastructure is a key factor in helping to ensure project success. How individuals communicate with other members of the TPO and the effectiveness of the media and communication tools

used is an important management concern throughout the entire life of a project.

- Chapter 3 – *Trust*. Trust and distrust play a central role in all aspects of our lives, from personal relationships to our daily interaction with work colleagues. In TPOs trust is initially based on our expectations and stereotypical perceptions of others, which is termed swift trust. As we interact with other project members our initial perceptions are challenged or reinforced and thus we start to develop trust (or distrust) based on how people behave. This complex area is influenced by our personal values and the values of the organisations that employ us. It is also influenced by our personal and corporate ethics.
- Chapter 4 – *Discussions*. Meetings and workshops are familiar forums for bringing project participants together (both physically and virtually) to develop relationships, share information and make informed decisions. Meetings are used for a wide range of purposes, but need to be used sparingly and be managed effectively to ensure they retain their relevance to the project. Workshops are usually convened for a specific function, for example to assist with team building, and these also need to be used sparingly and facilitated professionally to remain effective.
- Chapter 5 – *Decisions*. Projects involve a considerable amount of problem identification, problem solving and definition. Some of this is done in isolation, but most is conducted in disciplinary and multidisciplinary groups, thus the ability to make decisions is influenced by group dynamics, the effectiveness of communication within and between groups, and the level of trust that exists between stakeholders. Group decision-making techniques, the ability to recognise and hence avoid groupthink and the ability to deal with design changes are addressed in this chapter.
- Chapter 6 – *Context*. This chapter deals with a theme running through the book, that of context. Before the TPO is assembled it is crucial that the contextual factors for a specific project are investigated, analysed and understood. The goal is to assemble the most appropriate organisations and individuals for a specific task, and this can be achieved only once the client, the physical and the social contexts have been addressed. The ability of managers to demonstrate leadership and implement the most suitable process management model to the context will also be influential in determining project success.
- Chapter 7 – *Assembly*. Given an appreciation of the fundamental factors and the project context it is now possible to consider procurement and the assembly of the TPO. Assembly is often rushed into without adequate consideration for the consequences of the interfaces created, either by design or accident. The argument here is that taking time to put the right individuals and organisations together before the project starts will be beneficial to the effectiveness of the project and stakeholders alike in the long run.

- Chapter 8 – *Development*. Assembling the most appropriate people and organisations to work on the project is only part of the challenge. Once individuals and groups start to work on the project, or more specifically on their particular work packages, relationships start to develop and evolve. The development of an effective TPO cannot be left to chance and considerable effort is required to keep all participants working effectively and efficiently. This requires an understanding of group development, inconsistent project membership and conflict management.
- Chapter 9 – *Learning*. Continual improvement and the ability to learn from a variety of sources is the theme of this chapter. The opportunities and challenges of learning during the project and after project closure are explored, alongside the opportunity to reflect on, and learn from, experience. Storytelling as an aid to learning is also explored.
- Chapter 10 – *Implementation*. As a means of illustrating the practical application of some of the ideas contained within the book a case study is presented. The content is based on a longitudinal research project that investigated the development of a people-centred method for managing AEC projects. Facilitating face–to–face interaction within multidisciplinary groups of stakeholders is fundamental to the approach of the case study organisation.

Given the deliberately generic content of the material it is intended that the book will be relevant to a wide range of project types and approaches and will be of value to a wide range of project stakeholders. In addressing a wide readership it has not been possible to address specific forms of contract, or indeed the legal implications of how interdisciplinary construction project organisations are constituted. These are important considerations, but they are specific to the context of the reader and their organisation's business objectives. Nor is there any description of standard project management techniques and tools, as these are adequately covered elsewhere. A conscious decision has also been taken to present a balanced view and not to be swayed by topical management fashion for certain types of interdisciplinary arrangements, as the one thing we can be sure of is that this will change with time.

1 Interfaces

A discipline is a branch of learning or instruction. Synonyms for the word 'discipline' include domain, field, profession, regime, subject and trade; implying a certain uniqueness or specialism. Within the AEC sector there is a wide range of established specialisms, or disciplines, which are represented by professional bodies and trade organisations. Examples of professionals are architects, engineers and surveyors; examples of tradespeople being electricians, plumbers and masons. Whatever our chosen discipline, as individuals we undertake a programme of education and/or training that is accredited by a professional institution or trade body and designed to develop subject-specific knowledge. It is through learning and training that we are exposed to unique ways of doing things, unique ways of communicating, unique ways of behaving; all of which encompass the values of our chosen discipline. Once we have successfully completed our education and/or training a final examination and interview is required to ensure that we can demonstrate the necessary knowledge to complete our rites of passage and become a member of our chosen discipline. Once we are accepted into a profession or trade our disciplinary values are further developed and reinforced through working on projects and engaging in lifelong learning activities. Taking a random sample of a profession we would expect to find differences in interests, skills, experience and knowledge gained from working on projects, but a reassuring consistency in terms of educational, professional and disciplinary values. Although we need to be careful in resorting to stereotypes, we should have a reasonable idea of what an architect, engineer or surveyor can add to a project.

Grouping together individuals from one domain (for example architecture) is termed a disciplinary group, team or organisation. A small group of architects working together as a professional partnership would be classified as a disciplinary organisation. Indeed, it is not uncommon for professionals such as architects, engineers and surveyors to work with their peers in small groups comprising five or fewer professionals. By doing this the organisation can provide architectural, engineering or surveying services to its clients. Interaction within the organisation is with people who share a common education, language and professional values. Interaction with individuals from complementary disciplines and trades will take place through individual projects, in which a wide variety of disciplines interface to realise the project goal(s).

Over a long period of time disciplines develop unique subject knowledge, specific ways of working and codified ways of communicating – a unique culture. This uniqueness serves to both identify the special characteristics of the discipline and develop stereotypes, while also helping to reinforce and protect the profession's knowledge domain. Boundaries to a specific discipline are established and defended through education, practice and the actions of professional and trade bodies in promoting and defending their specialisms. Seen from a business perspective, it is concerned with establishing and growing market share for professional services. Seen from the perspective of the profession it is about expanding the knowledge base. Paradoxically, although the development of a discipline is a strength, it can make it difficult for those positioned outside the field to access the knowledge held within, or to fully understand the peculiarities of bespoke working practices and the language used. This has sometimes led commentators located outside the profession to conclude that professionals are arrogant and stubborn. Disciplines also run the risk of becoming so specialised that they become detached from other specialists and hence vulnerable to market forces.

Interdisciplinary

The word 'interdisciplinary' encompasses more than one branch of learning or instruction: the interface of different domains. Combining the disciplines of architecture and engineering creates an interdisciplinary subject in education, architectural engineering, and a hybrid professional, the architectural engineer.

Grouping together individuals from more than two disciplines is known as a multidisciplinary group, team or organisation. In the context of the AEC sector, a multidisciplinary project group or team could, for example, include an architect, an architectural technologist, an architectural engineer, a structural engineer, a services engineer, a quantity surveyor and a project manager. Bringing together different knowledge domains within the same organisation allows the business to offer a wider range of expertise compared with the disciplinary organisation. This means not that a multidisciplinary office is better or worse than the disciplinary one, simply that the two organisations are able to offer different services and so occupy slightly different market segments.

In contrast to small organisations, the majority of large businesses are multidisciplinary organisations because a wide variety of technical and managerial skills are necessary for the business to function effectively and efficiently. Some organisations may have a heavy bias towards one domain, for example engineering, or they may be composed to create a balanced interdisciplinary organisation that is capable of contributing to a wider variety of project types, often providing their clients with a 'one-stop-shop' for expertise.

The term 'interdisciplinary working' is used to describe the interaction of individuals from different disciplines, both within organisations and within temporary project organisations. With increased differentiation within the AEC sector has come a greater need for integrated working. Interdisciplinary ideologies such as constructability (buildability) and project partnering are founded on the concept of sharing disciplinary knowledge and communicating across disciplinary boundaries to improve product delivery.

Buildings and civil engineering works are complex undertakings, which require the coordination of technological, design and management expertise if they are to be realised. Even the most simple project will require a diverse assembly of people to guide it from inception and development through to successful completion. On small projects, for example an extension to a house, this collection of individuals may form a small team that is relatively stable throughout the life of the project. When we move to larger, more demanding buildings, we find that projects are composed of many groups of individuals with complementary knowledge, drawn from a wide cross-section of disciplines and organisations, increasingly working across international boundaries and composed of individuals from many different cultures and nations. Temporarily bringing these parties together to work on an AEC project forms a loosely coupled, dynamic, multidisciplinary project coalition. Typically these project coalitions are best described as temporary project organisations or temporary multiorganisations. It is, however, still common to see the term 'project team' used in the literature and in everyday conversation, which can give the misleading impression of a cohesive entity, which it is not. For clarity and consistency the term 'temporary project organisation' is used throughout this book to describe the assorted and transient collection of organisations and individuals engaged in an AEC project.

Fundamentals

Interdisciplinary (or interorganisational) working is concerned with the manner in which individuals from a variety of disciplines interact during the life of a project, i.e. it is concerned with the effectiveness of the various temporary interfaces. As already noted, interdisciplinary working is usually achieved through a coalition of disciplinary groups and teams and multidisciplinary groups and teams, by means of a TPO. These dynamic organisations exist for the sole purpose of delivering a project, and comprise a loose coalition of multi-skilled individuals with varying values, attitudes and goals. A small number of government publications have been highly influential in bringing about greater awareness and subsequent attention to the importance of effective TPOs in the UK. Two publications, *Trust and Money* (Latham, 1993) and *Constructing the Team* (Latham, 1994), helped to highlight the issues surrounding teamwork and project partnering. The fundamental tenets of construction projects were subsequently reinforced in *Rethinking Construction: The Report of the Construction Task Force* (Egan, 1998) and *Rethinking Construction: Accelerating Change* (Egan, 2002). These reports aimed to bring about a change in attitude from an adversarial and fragmented sector to one that is more trusting and better integrated. The publications express similar sentiments to those contained in earlier reports by Simon (1944), the Phillips Report (1950) and the Emmerson Report (1962), which also argued for better communication and more effective interaction between project participants. In many respects the underlying message has not changed, but the context, technologies and language have.

Inadequacies reported in the government publications appear to stem from poor interaction practices during the life of the project (Emmitt and Gorse, 2003, 2007). Interaction affects the strength of the relationships between the actors and ultimately colours their ability to transfer knowledge and appropriate task-based information to complete projects successfully. Team building, the discussion and subsequent sharing of values, resolution of minor differences and conflicts, question asking and the creation of trust are just a few of the factors that are crucial to the smooth running of projects and are reliant on the ability of the actors to communicate effectively and efficiently. It follows that the interaction of individuals and organisations should be the primary concern of those charged with managing projects.

The government-led reports have inspired many books, reports and articles that present a very positive argument for relational (interdisciplinary and collaborative) forms of working; examples being Baden Hellard's *Project Partnering: Principle and Practice* (1995) and *Trusting the Team* by Bennett and Jayes (1995). The message is that the AEC sector needs to move from 'segregated' teams to 'integrated' teams to improve performance and hence deliver better value to customers and the users of buildings. Although there has been an increase in the number of AEC projects that use relational forms of contracting (such as project partnering), it is still relatively modest compared with the more traditional approaches (such as competitive tendering). The majority of projects are still conducted in ways that, on the surface at least, are based on distrust and non-integrated or isolated working. This does not necessarily mean that the majority of projects are less efficient or less effective than those conducted in a spirit of trust and collaboration; it simply means that the project philosophy, i.e. attitudes of the participants, is different. Applied research into how individuals interact within AEC projects (see Emmitt and Gorse, 2007) indicates that the issues relating to project work are not simple or straightforward. For a balanced approach it is necessary to consider the fundamental characteristics of all types of interdisciplinary project, be they collaborative, competitive, integrated or fragmented.

Temporary project organisations

The way in which people from different domains are brought together and how they interface with other disciplines to realise project goals in an effective and efficient manner is a concern to the sponsors of projects, who desire value from their investment. Similarly, this is also a concern to those participating, who need to demonstrate value to their clients and make a reasonable profit from their contribution. TPOs represent short-term business relationships involving a variety of organisations and individuals with complementary skills, but with varying business objectives. The overall performance of the TPO will be determined by the collective effectiveness of the contributors, i.e. their ability to work with others towards a common goal.

Successful realisation of the project is dependent upon many inter-related factors, which tend to relate to the management of the process and people issues. Project managers need to establish the most appropriate processes for a project and engage the most suitable organisations and individuals. This calls for an understanding of many complementary domains and how participants from these domains are likely to interact. Participants need to be able to work with the agreed processes and the communication technologies employed for the project; they also quickly need to be able to establish ways of working with their new colleagues. Taken at face value this would appear to be a relatively simple thing to do, although the reality is a little more complex given the characteristics of AEC projects.

Temporary project organisations are set up for the life of a project and disbanded on successful completion of the building (or at specific stages of some large and complex projects). This usually creates new relationships for each project, a temporary social system, which provides the project manager with the immediate challenge of team building, establishing open communications and developing a level of trust between the project participants as quickly as possible. Many suppliers work across different sectors, for example the manufacturers of an energy-saving paint also produce skin products for the cosmetic industry. With a few exceptions of repeat building types and clients with very large property portfolios there are few established or stable supply chains, unlike, for example, the car industry. Interfaces within projects are rarely stable and with the exception of some repeat projects supply chains are unique to each project. In situations in which the organisations have worked together previously, for example on repeat projects or in strategic alliances, the individuals involved in the project are not always the same as those involved in previous projects, and so team building is still necessary to develop relationships at the interfaces. Further challenges relate to the timely exchange of accurate information and managing the web of inter-related and interdependent activities necessary to achieve project completion.

Concomitant with all forms of human interaction, AEC projects will be prone to difficulties as relationships form and evolve during the life of the project. Individuals may find themselves working with others whom they perceive to be less than trustworthy or whom they simply do not like, and it would be foolish to think that incompatibility will not influence the effectiveness of the project. It is also inevitable that there will be a constant and creative tension between the design and the realisation of buildings as the TPO collectively strives to deliver value to the sponsor of the building project (the client), generate value for its own organisations (and shareholders) and in a wider sense demonstrate value to society (both current and future generations). Successful design managers and project managers are able to manage relationships so that the positive aspects of interaction are encouraged and the negative aspects dealt with quickly so as not to undermine the effectiveness of the TPO. Whatever the approach taken to the assembly and development of a TPO the ultimate aim is to create and maintain an effective and efficient coalition that delivers a final product that meets all performance parameters. Good managers tend to be sensitive to the interactions between individuals, disciplines and organisations during the life of a project and are able to anticipate challenges and hence mitigate the possibility of incompatibility and thus reduce unnecessary negative conflict. To do so requires a thorough understanding of the role of project stakeholders, participation, collaborative working and collaborative technologies.

Project stakeholders

The word 'stakeholder' is used to describe a person, group or organisation with an interest, a stake, in a project. A stake in the project may be financial, contractual and/or emotional. A project stakeholder is able to affect, or is affected by, the combined decisions and actions of the temporary project organisation. Taking this as our definition it follows that everyone contributing to the project and those interacting with the completed building are stakeholders. It is, however, often necessary to be specific about who the stakeholders are and how they can affect the project and also how they can be affected by it. In its widest sense a project stakeholder may be someone who will use the building, but who has little or no input to the project decision-making processes; but they will be affected by the outcome of the project. Other stakeholders may have very little to do with the project but will have a financial stake, examples being investors and insurers.

Project stakeholders have different levels of responsibility for the project. Some are linked to the project through some form of contract (e.g. for architectural services), some by their statutory duty (e.g. local authority town planners), some simply by the fact that a building is going to be erected in their locale (e.g. a neighbourhood pressure group). When planning projects it can be useful to map and classify stakeholders in terms of their contribution to a project, for example contractual, non-contractual, financial, non-financial, from which it is possible to clarify their decision-making responsibilities. Stakeholder mapping can also help to clarify relationships, identify communication routes and establish the most important contributors at specific stages in a project. This is important because stakeholder engagement is related to the level of participation in a project.

Participation in projects

Participation is the act of sharing or taking part in group decision-making processes. In the context of a project participative processes should include a wide range of stakeholders, from the building sponsor and the professionals working on the project through to the building users and representatives of the local community in which the building is located. The intention of a participatory process is to achieve a higher degree of synergy by bringing multidisciplinary actors together to share their knowledge, hopefully resulting in an outcome that would not be possible working individually. Outcomes are by consensus and the group members share ownership of 'their' decisions. Sometimes actors have equal participation rights, although it is more common for participants to have different levels of responsibility within the project. This is addressed in more detail in Chapter 5.

Participation can occur synchronously (at the same time, e.g. within a meeting forum) and asynchronously (not at the same time, e.g. by an exchange of letters or emails). Participation can be facilitated by online forums and interactive media and for large projects with many stakeholders, for example a new hospital, ICTs provide a convenient tool for allowing a large number of people to participate and express their opinions.

Within a project setting it is important to communicate using a language that all participants can understand, to be able to deal with conflicting views

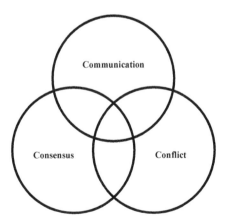

Figure 1.1 Fundamentals of participation.

in a positive manner, and to try to reach consensus within a given set of project parameters (Figure 1.1):

- *Communication.* Participation will be more effective if individuals are able to understand others and communicate effectively. This often means that professionals need to choose their words and visual media carefully when interacting with non-professionals and other disciplines. A common language, or at least a simplified language free from acronyms and technical terminology, should help to encourage participation and prevent individuals from feeling excluded from discussions. Similarly, simplified drawings and diagrams can be more effective in getting a message to laypersons than technical drawings. This applies to face-to-face discussions, where there is a chance to ask for clarification, and to online participation, where the opportunity to ask for clarification may be less convenient.
- *Conflict.* In an ideal world participants will be selected based on their expertise and knowledge and their ability to interact harmoniously with other participants. This is rarely the case, as individuals are often allocated to projects by line managers based on their availability, not their personality. Resource allocation policies of an individual organisation will determine the individuals who are assigned to a particular project. The individual may have little or no say in the matter. Not surprisingly, when a number of individuals representing their organisations interact on the project there is always the potential for incompatibility and negative conflict. Participation by representatives of local pressure groups may be confrontational and deliberately disruptive when a proposed development is not welcome in their neighbourhood. Participants must be prepared to address differences of opinion so that they are able to deal with conflict in a positive manner in an attempt to reach consensus.
- *Consensus.* Participants should be working towards decisions made by consensus, which calls for tact, diplomacy and the ability to negotiate. Sometimes it may be necessary to 'agree to disagree' so that the project can proceed and not be stifled by an impasse. This means that there needs to be relatively open communication to allow informed discussion and a supportive environment in which participants feel comfortable to participate in the discussions and the decision-making process. Clarity of communications and transparent decision making are also fundamental to achieving consensus.

Participation levels will vary throughout the various stages of a project and managers need to appreciate, and hence anticipate, the influence of participation on the progress and eventual outcome of the project.

Collaborative working and teamwork

The word 'collaboration' (or 'collaborative') is sometimes incorrectly used as substitute for 'participation' (or 'participative') and vice versa. Participation usually involves various levels of responsibility and power, thus the interaction of some participants will carry more weight than that of others. Collaborative processes are concerned with equal participation, equitable power and shared decision-making responsibilities. Given the inequality of participation rights within projects the term 'collaborative' may be a little misleading; however, the use of the terms 'collaboration' and 'collaborative' have become widespread in the AEC sector.

Collaboration refers to cooperation with others, the uniting of labour to achieve a common objective. A collaborative project is one that involves multiple actors who work together and hence are mutually dependent upon one another. Collaborators should be prepared to listen to others, treat their ideas with respect and give each actor equal decision-making power. The aim is to resolve problems more effectively and also produce better outcomes than those likely through non-collaborative approaches. This often means that individuals and the organisations that they represent may have to relinquish power; this can be difficult for some participants to deal with, especially if they are more orientated towards a conflict-based approach to business. It follows that it is useful to recognise that collaborative processes may not suit everyone, thus care is required when selecting the members of the temporary project organisation.

The architects John Ruskin and William Morris both promoted collaboration between architects and tradesmen, and Corbusier argued for closer collaboration between architects and engineers. The Bauhaus movement also pursued a collaborative ideal in its earlier years. In particular, Walter Gropius was devoted to the notion of collaboration between the fields of design and architecture, eventually setting up the Architects' Collaborative in 1945 (Gropius and Harkness, 1966). Teamwork has also been promoted by architects of a more entrepreneurial disposition. For example, *Architecture by Team* (Caudill, 1971) is notable for the language used and the author's passion for an interdisciplinary approach. Caudill dismisses the notion of individualism in favour of interdisciplinary teams, describing a number of different approaches to group practice and claiming that in the future 'great architects will be on great interdisciplinary teams' (Caudill, 1971: 31). Caudill's book and similar publications on architectural management reflect the realities of architectural practice, and tend not to sit easily alongside the more heavily promoted stereotypical image of the architect as an individualist and non-team player (see Saint, 1983). Publications by Gutman (1988), Cuff (1991) and Emmitt *et al.* (2009) help to demonstrate the collaborative, inclusive nature of architectural practice. Collectively these are important works in helping to promote and develop an understanding of interdisciplinary teamwork in AEC, a philosophy practised by many architectural and multidisciplinary consultancies over the years.

Collaborative technologies

Information technologies have transformed, and continue to transform, the way in which buildings are designed, manufactured, assembled and used. Improvements in the visualisation of designs, modelling and communication between the participants have helped to provide a better understanding of the management of intricate processes. In particular the development of IT and ICT such as project websites has made it easier to work concurrently and collaboratively from remote locations. Four-dimensional computer-aided design (CAD) models and building information modelling (BIM) provide the means for addressing the fourth dimension, time. These virtual models provide the interconnection between design information and the planning and scheduling activities (three-dimensional + time), providing animations of construction sequences. BIM provides users, regardless of their physical location, with the opportunity of testing, revising, rejecting and accepting design ideas in real time, i.e. it provides the means for collaborative design. BIM also provides a tool for improving the efficiency of communication within the TPO, as it is able to handle the vast amount of information required for coordination. Technical concerns over interoperability of various software packages and the availability of bandwidth to allow the large volume of data traffic to flow smoothly are ongoing concerns, but are being addressed. There is a perception, promoted by the vendors of the software, that better software will lead to better designs and better management of projects, although there appears to be dissonance between what the vendors claim and what happens in practice (Otter and Emmitt, 2007). We need to remember, for the time being at least, that people, not software, manage interdisciplinary projects.

Interdisciplinary project teams and groups

The overall message and approach reflects our current preoccupation with improving the effectiveness of the TPO through better interaction practices. This is particularly pertinent when considering the challenge of assembling a TPO to work across international boundaries. Interdisciplinary TPOs are composed of specialists operating in a disaggregated sector, each carrying different (one would hope complementary) values and intentions from other actors. This characteristic is common to all AEC projects, regardless of the type of project and the procurement route used. It is convenient and common to use terms such as 'project team' and 'construction project team' to encompass all those contributing to an AEC project. The danger in using such terms is that the impression given is one of a homogeneous and stable organisation; which it is not. Putting an assortment of individuals together and calling them a project team is no guarantee that they will be able to interact effectively and efficiently. The danger is that the underlying principles of how groups and teams function are not adequately discussed and understood before embarking on a new venture.

In large and/or complex projects the number of people involved can be extensive, each bringing expertise, values, aspirations and prejudices to the project. Fortunately, from a manager's perspective, the large number of contributors tend to be centred in small groups and teams. It is the field of social psychology that has addressed the complexities of how groups and teams function. Within this body of literature it is evident that there are a number of challenges associated with the development of effective work groups and teams; indeed there are conflicting views as to how to achieve effective group and team performance. It is well established that successful projects tend to be characterised by effective communication between project participants and their coordinated (harmonious) activities.

Teams

The definition of a team is a small group of individuals (three or more) working towards a common goal within an organisation. Small teams can also be found working across organisational and departmental boundaries, sometimes referred to as 'cross-functional' teams. Teams usually comprise a mixture of individuals who take on different roles in order to make the team effective. Meredith Belbin's (1981) work on management teams is most widely known in this area, being derived from the observation of groups, and providing a comprehensive analysis of team roles. Belbin's message is that group effectiveness is determined by the successful coordination of specific roles. This means that the team members need to be carefully selected, using the Belbin questionnaire (or other questionnaire-based schemes), so that specific roles are addressed.

Interdisciplinary, cross-organisational project teams differ from teams assembled within an organisation. Contributors are usually chosen on their technical knowledge and interpersonal and problem-solving skills, not on their ability to take on a specific role within a team, which is often a secondary concern. Indeed, given the nature of TPOs it can be very difficult to identify specific team roles within the project, which is different from an organisational context in which relationships and roles tend to be more stable.

With the advent of electronic communications it has become common to define teams (and groups) by their location and use of electronic media. Although the terminology tends to vary, there are three terms in common use: co-located, virtual and global teams (Gameson *et al.*, 2005):

- *Co-located teams.* These are teams located in the same physical (office) space. These could be disciplinary or multidisciplinary groupings of individuals. It is highly unlikely that the project team will constitute one large co-located team. Usually AEC projects are characterised by a number of smaller co-located teams (both disciplinary and multidisciplinary) that exchange communications electronically and meet occasionally (e.g. at progress meetings).

- *Virtual teams*. These teams are formed with a mix of co-located people and people connected by electronic networks, for example a number of consulting offices located at various geographical locations within the UK (who might meet occasionally) with work packages outsourced to organisations located overseas.
- *Global teams*. These are teams that do not meet physically and might never meet during the course of the project. This is common on large international projects, but could equally apply to any size of project. In these types of relationships the performance of the TPO will be influenced by the technologies being used and the abilities of the participants to use them.

It is highly probable that in many AEC projects all three types of team will exist at some stage in the project life cycle. Their advantages and disadvantages need to be addressed for a particular project context. Defining teams (and groups) by their geographical location and use of ICTs may be helpful in establishing appropriate project communications. It may also be useful to define TPOs by their geographical scope of work (in terms of their culture and language), as local, national and international:

- *Local*. Many small to medium-sized organisations operate in a relatively small geographical area. They tend to know the other organisations and share a similar language and culture.
- *National*. Some organisations restrict their work to their mother country, often reluctant to engage in international projects because of concerns over unfamiliar language, culture, laws and competition.
- *International*. Organisations are increasingly working on international projects, collaborating with other international organisations to bring expertise to complex and challenging projects, aided by advances in ICTs. Ability to communicate in a language other than one's mother tongue may be necessary and an ability to understand unfamiliar cultures, laws and working practices is a prerequisite for success.

Groups

The focus in this book is on work groups and how they function effectively within the context of a TPO. Work groups are characterised by the need to undertake a task (or a number of tasks), usually within a well-defined time frame (e.g. Tajfel and Fraser, 1978). Work groups are termed 'task-orientated' because the group members should be working together to achieve their task(s). Achievement of a group's goals depends on concerted action and so group members must reach some degree of consensus on acceptable task and socio-emotional behaviour before they can act together (Hare, 1976). This type of interaction has been described as task specific and social. The social element of interaction is developed through emotional exchanges that are used to express a level of commitment to the task and other members. To accomplish group tasks relationships need to be developed and maintained. The level of interaction associated with maintaining, building, threatening and breaking up relationships will be a function of socio-emotional interaction and will be subject to group norms (discussed below).

There is a difference between disciplinary groups (groups comprising members of the same profession) and multidisciplinary groups (groups comprising members from a variety of professions). In disciplinary groups the objectives and values of each individual are likely to be similar to those of other members. In multidisciplinary groups there is likely to be larger variation in objectives and values, thus interaction and compatibility of group and individual goals in multidisciplinary groups is fundamentally different from the corresponding process in disciplinary groups (Cartwright and Zander, 1968; Lieberman *et al.*, 1969).

The size of the group is also worthy of consideration. Although AEC projects can involve a large number of people, it is common for specialists to operate in relatively small work groups or teams. Representatives for these groups and teams will usually come together and interact during project meetings and workshops, and given the complexity of construction projects this often involves a large number of people. However, much of the work is done outside these large gatherings by smaller groups of specialists, often working concurrently. The terminology used to describe these work groups is 'small groups' and the most effective size is somewhere around six or seven people. Professional differences between group members may make the establishment of team goals difficult to achieve. Reconciling individual objectives with the overall group objective may be equally problematic, resulting in ambiguity and in some cases intragroup conflict.

Integrated temporary project organisations

The word 'integration' is used to describe the intermixing of individuals, groups or teams who were previously segregated. Integration within TPOs can occur on a number of different levels, from seeing the whole project process as an integrated one, to viewing the concept as simply bringing together two separate work packages. The term 'integrated teams' has come into widespread use in the AEC sector; although tautological, it seems to be used to describe TPOs that are comparatively more integrated than might otherwise be the case. The majority of the literature promotes a highly positive view of integrated teams, while failing to acknowledge the inherent sociological and psychological challenges. For example, problems associated with 'groupthink' have yet to be adequately addressed as have related issues concerning positive and negative conflict. These are some of the issues that should be confronted if we are to approach the management of construction projects from an informed position.

Integrated design, supply and production processes are facilitated by cooperative interdisciplinary working arrangements. Integrated teams encompass the skills, knowledge and experience of a wide range of specialists, often working together as a virtual team from different physical locations. Multidisciplinary teams may be formed for one project only, or formed to work on consecutive projects. Although there has been a move towards more collaborative working arrangements based on the philosophy of project partnering and strategic alliances it is difficult to see evidence of real integration; instead there are pockets of collaborative work within and between projects.

Focusing on integrated processes is only part of the challenge. It is also necessary to look at the individuals involved with the project and to look at how integrated their contributions are. How and when, for example, are the contractor and main sub-contractors involved in the early design phases? Are they an integral part of the design decision-making process or are they merely invited to attend meetings and asked for their opinions? For real integration to work there needs to be social parity between actors, which means that professional arrogance, stereotypical views of professionals and issues of status have to be put to one side, or confronted. It also means that, in many cases, project teams need to be restructured and the project culture redefined through the early discussion of values, for example by value management exercises.

Integration of design and construction activities can achieve significant benefits for all project stakeholders. Improvements in the quality of the service provided and the quality of the completed project, reduced programme duration, reduced costs, improved value and improved profits are some of the benefits. Traditional procurement practices are known to perpetuate adversarial behaviour and tend to have a negative impact on the product development process and hence the project outcomes. Focus has tended to be on limiting exposure to risk and avoiding blame at the expense of creativity and innovation. The creation and maintenance of dynamic and integrated teams is a challenge in such a risk-adverse environment. Fostering collaboration and learning within the project frame requires a more integrated approach in which all stakeholders accept responsibility for their collective actions. Project partnering is one approach that can help to bring the actors together, which, if combined with value management techniques, can provide both the philosophy and the tools to effectively develop an integrated TPO. Using new technologies and new approaches, such as off-site production, is another approach, which significantly changes relationships.

Independent working

One of the main arguments in promoting integrated and collaborative forms of working is that the traditional approaches to construction are character-ised by fragmented working practices, insular cultures and conflict. Emotive terms such as 'silo mentality' and 'over the wall' are used to enhance the argument for doing things differently. Unfortunately, such arguments tend to be based on an overly simplistic understanding of how professionals work. Even with independent working there will be some interaction between work packages, and, although the process plan may show clear separation, in reality there is likely to be a degree of informal communication, interaction and iteration between the work packages. Sometimes it is desirable or neces-sary to work with a certain amount of separation between work packages, remembering that the work is inter-related to other activities.

Fundamentals of group interaction

Research into group interaction has shown that the process is difficult to understand in its entirety (Poole and Hirokawa, 1996). Interactions of participants who are subject to psychological, social and contextual influ-ences make the subject difficult to research in live business settings. Much of the empirical research on group interaction and communication focuses on group satisfaction, which has been carried out in experimental groups, usually groups of students. These studies frequently deal with perceptions of how members feel after a particular group encounter or how they think they would feel under particular circumstances or situations. They provide a useful insight into individuals' perceptions of behaviour, but the work has not measured behaviour or performance. Indeed, perceptions of how peo-ple think they behave and how they actually behave can be quite different. Research into how individuals interact within temporary project organisa-tions is relatively sparse, a point made by Emmitt and Gorse (2007).

The length of time a group has existed is known to influence the perform-ance of the group. Newly formed groups are known to be relatively ineffective at first because a great deal of emotional energy is put into establishing relationships with other group members. Once the group is established less effort is needed on these tasks and so the group can focus on the task to hand and tends to become more effective over time. Unfortunately, a change in group membership effectively renders it a new group once again and hence the group becomes less effective until relationships have been established.

All groups will not behave in the same way; however, there are some fundamentals of group interaction that are common to small groups. Hartley (1997) has proposed an 'integrative model' to help people to start to think about how groups function. This simple yet effective model has three levels of analysis, each of which is needed to fully understand group interaction, namely context, surface behaviour and hidden agendas (Figure 1.2). These are now discussed in a project environment.

Context – social and cultural background

This includes the setting in which the group is operating as well as the social and cultural background of the group. The setting is particularly important in AEC projects and is discussed further in Chapter 6. Social and cultural factors concern issues such as pride in the group's achievements, group loyalty and values of cooperation. Cultural differences, for example individuals from countries with different value sets and/or individuals belonging to professions with different value sets, need to be explored, otherwise group members may find it difficult to interpret each other's behaviour (see Chapter 4). This can lead to misinterpretation, ineffective decisions and in the worst cases conflict within the group (which needs to be addressed if the group is to remain productive). In the case of AEC projects the group culture of the TPO is temporary and highly dynamic, morphing through the various stages of the project as specialists enter and withdraw from the project. Given this dynamic process it may be slightly misleading to refer to a 'project culture' in the singular. The social context will be shaped by the changing nature of interpersonal interaction over the project life cycle, which can be lengthy, and which tends to create a series of project cultures.

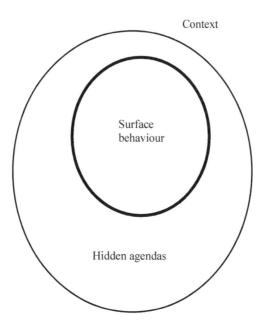

Figure 1.2 Context, surface behaviour and hidden agendas.

Surface behaviour – tasks and procedures

This level is concerned with the overt behaviour of the group. This includes the task to be completed and the procedures that the group uses to realise the task, i.e. how the members communicate and the roles that individuals take on. In a TPO many of the tasks and procedures will be conducted within the offices of the contributing organisations. It is when disciplines interact during, for example, project progress meetings that surface behaviour can be observed by the other contributors. Surface behaviour can be observed by other group members, project managers and researchers, providing an indication of how our colleagues act in a given situation.

Hidden agendas – the interpersonal underworld

In some respects this level is perhaps the most interesting and also the most challenging, simply because it is usually hidden from other group members, project managers and researchers. It is not unusual to hear professionals refer to the 'battles going on under the table', indicating that the surface behaviour is not communicating the entire picture. In the interpersonal world we are concerned with the emotional aspects of group interaction, the conscious and subconscious issues including values, trust, likes, dislikes, attitudes, opinions and perceptions of others. According to Hartley (1997) this is the one aspect that everyday work groups will avoid discussing openly. Not only does this make it very difficult to research interaction with groups in their natural setting, but also it can have a detrimental effect on the ability of the group to develop. Reluctance of group members to engage in discussion about underlying issues, and hence bring them to the surface, can often be a factor in poor performance. Facilitated workshops are one way of trying to tease out hidden agendas in a temporary project organisation (see Chapter 4). Unfortunately, hidden agendas (such as level of commitment and purpose for participating) tend to surface during a crisis, often making a difficult situation much worse.

Recognising and managing interfaces

Bringing different disciplines together helps to ensure that the various aspects of a project are developed in a manner befitting the client's aspirations. However, this is rarely an easy task because, even if the organisations have worked together previously, it is unlikely that the individuals involved will all know one another, and so the early phases of interaction are concerned with exploring and learning how to work effectively with new people from unfamiliar organisations. Creating a TPO also creates a vast number of interfaces between organisations, organisational departments, teams, groups and individuals. Some of these interfaces will be new, some may be re-established after a (short) period of not working together and some may be continued from previous projects. Effort expended in forming relationships, testing the abilities of new acquaintances and trying to establish their trustworthiness at a very early stage in the project can be beneficial in quickly establishing effective working patterns. Trying to establish compatibility as early as possible in the life of a project can be a time-consuming task, but this effort can have significant dividends in terms of quickly building effective working relationships, developing mutual trust and reducing the likelihood of misunderstandings and unnecessary conflict (Emmitt and Christoffersen, 2009).

Recognising interfaces

One of the most important considerations for managers is the interface between the different organisations, groups and teams contributing to a project. Interfaces between construction materials form physical boundaries or joints, which are relatively easy to design and manipulate to achieve the required technical performance for a building because they can be seen. Interfaces between project organisations form softer boundaries that are determined by legal contracts (which specify responsibilities), but which are often blurred as individuals work informally with others at the margins of the boundaries. Boundary conditions are shaped by organisational cultures, business goals, tools and techniques (i.e. interaction of people and processes).

Organisational and cultural boundaries within construction projects are constantly changing: individuals enter and leave the TPO at certain stages (i.e. they have separate goals) and the team changes in size and format. Boundaries of responsibility and interests in the project are constantly in a state of flux. Obvious boundaries are the interfaces between client and brief-taker; brief-taker and design team; design team and contractor; contractor and sub-contractors. Other more subtle boundaries, for example between architects and engineers, also exist. In many cases the interface is effective and projects progress well, delivering a building that satisfies both client and end-users. Conversely, interaction difficulties may be experienced and relationships can deteriorate and quickly spiral out of control into a dispute, which rarely benefits anyone.

Interface management

Interface management is concerned with the relationships between interdependent organisations working towards a common goal (Wren, 1967). Given that the number of interfaces can be significant within a TPO it may be useful to consider two fundamental types of interface that are constant throughout the project:

- *Organisational (business) interfaces*. Organisational interfaces are mainly defined by contracts and the project context. Interorganisational relationships are concerned with organisational culture and the interoperability of management and ICT systems. Although the relationships can be dynamic, they are relatively straightforward to define, map and manage through the life of the project.

- *Personal interfaces.* Individuals interface with others representing other organisations, not the organisation *per se*, thus interfaces are coloured by the ability to communicate and work with representatives of other organisations. The effectiveness of the relationships is dependent on the compatibility of the individuals concerned. These interfaces are more challenging to define, map and manage because over the course of a project it is not uncommon for individuals to move jobs or be allocated to different projects, thus introducing new individuals and interfaces.

Interface management is a particular concern of the project manager and increasingly the construction design manager, both roles being central to the coordination of information and resources and a feature of supply chain management. Some of the main areas to consider are:

- *Interface definition.* Map primary and secondary interfaces and identify areas of uncertainty.
- *Responsibilities.* Clearly defined and visible areas of responsibility can help to reduce disagreements and disputes.
- *Communication.* Clear and effective communication is central to interface management.

These areas should be considered both at the early planning stage of a project and on a regular basis as the project evolves because the interfaces will change over time. Although this may seem a daunting thing to do, ICTs can make the task much easier because the formal communication traffic can be analysed.

Practical challenges

One might be forgiven for asking how it is possible to plan and subsequently manage a project without making sufficient effort to understand how the participants are likely to interact. Simply applying a standard way of working (a standard process plan for example) to every project may result in ineffective and hence inefficient interaction between the contributors, simply because the contributors are uncomfortable with the way in which they are being asked to work. Having stated this it is not uncommon for managers to take this approach, applying the same way of working from one project to the next (even when it might not be appropriate to do so). Clearly the lack of time to consider alternatives is often a determining factor, but so too is the comfort zone of the manager and their penchant for certain approaches that work for them. Taking different approaches on consecutive projects poses risks if insufficient time for learning and gathering sufficient knowledge is not factored into project programmes. The challenge for managers (time and willingness to embrace change permitting) is to find an appropriate fit for the contributing organisations within a project context. In many respects this is a chicken and egg problem. Should the project processes be designed and the appropriate contributors selected to fit the processes, or should the key participants be selected and then the processes collectively discussed and agreed? There is no correct answer to this question; what is more crucial is that both processes and people are given equal consideration by managers.

End of chapter exercises

- List the advantages and disadvantages of assembling a TPO comprising (1) five members from the same discipline and (2) five members from different disciplines.
- Identify the main interfaces between the key disciplines/organisations during the life cycle of a project, for example a new school building in the UK. How would these interfaces differ if the school was to be constructed using (1) on-site construction techniques and (2) off-site construction techniques?
- If the school project was to be located in, for example, Portugal, with a TPO composed of organisations based in other European countries (e.g. French architects, Danish engineers, British project managers, Dutch prime contractor), how would this influence the interfaces within the TPO? List five key management challenges with this international TPO.

Further reading

Additional material on managing construction projects can be found in Walker (2007) *Project Management in Construction.*

2 Communication

Now that some of the basics of how project participants interact have been addressed it is appropriate to consider how we communicate with others in a project environment. Communication is a familiar and essential characteristic of human societies, underpinning our personal and business relationships. Communication is also of vital importance to everyone involved in, and influenced by, AEC projects. To interact in a project environment it is necessary to communicate with individuals representing a variety of organisations and interests. It is the dynamic interaction of people within temporary project organisations that constitutes a process of communication. Participants need to collaborate, share, collate and integrate significant amounts of information and knowledge to realise project objectives. To do this well requires a combination of good management, committed people, supporting ICT networks, interactive media and the opportunity to communicate effectively with other participants. It is of little use being knowledgeable about a particular profession or trade if we are unable to articulate and communicate our ideas to others. Key management competences of leadership and decision making are founded on competent communication skills. Indeed, without timely, accurate and meaningful communication the TPO cannot succeed in realising its objectives.

Effectiveness of communication is a significant factor in the successful completion of projects (Gorse and Emmitt, 2009) and is becoming increasingly important because of the growing technical and organisational complexity of AEC projects. Over the years a number of UK government-led reports have consistently drawn attention to the difficulties caused by the organisational systems through which construction projects are conceived and delivered (e.g. Phillips Report, 1950; Emmerson Report, 1962; Latham, 1994; Egan, 1998, 2002). Many of the challenges frequently reported have been attributed to the separation of design and realisation activities. It is claimed that separation of activities results in poor understanding within and between participants as well as ineffective communication between them. This has led to significant efforts to develop more integrated TPOs, which have been helped by the rapid development of ICTs. Combined with a better awareness of how individuals interact in a project environment, ICTs provide the means to better integrate work, helping to improve the effectiveness of communication and ease the burden of coordination. Care needs to be taken to ensure that there is no incompatibility between the users and the software, because this will hinder productivity (Otter and Emmitt, 2007).

Development and use of ICTs has been seen as one way of improving the performance of design and construction teams (e.g. Love *et al.*, 2001), becoming synonymous with the better integration of project participants (e.g. Waring and Wainwright, 2000). ICTs are also a means of improving collaborative working, although this is something that is proving to be elusive in practice (Damodaran and Shelbourn, 2006). The term 'collaborative communication technologies' (CCTs) has come into use to signify more inclusive and more accessible technologies for realising project work. Parallel to the development of technologies is a growing recognition of the need to understand the needs of individuals and how they communicate within temporary project organisations (Emmitt and Gorse, 2007).

Research has consistently shown that poor communication is the root cause of many problems in both organisations and projects. Failure to consult with, and inadequate feedback to, other members of the project coalition are among the primary causes of construction defects. Ineffective communication will have a detrimental effect on the performance of the project as a whole, hindering the development of efficient working relationships and in the worst cases leading to confusion, errors and wasted effort.

Fundamentals

Communication, as highlighted by Dainty *et al.* (2006), is a complex term, which can mean different things in different contexts. A survey undertaken by Dance and Larson in 1972 found more than 100 definitions of the term 'communication', since which time the number of definitions has increased (Trenholm and Jenson, 1995). Communication is one of those human activities that everyone recognises, but few can define satisfactorily (Fiske, 1990). This is particularly true of multidisciplinary TPOs, in which disciplines have their own understanding and perceptions of what it means to communicate. When architects and authors of architectural publications talk about communication they are frequently found to be describing the way in which their work (both their drawings and the completed building) communicates with the reader. Architects talk of 'reading' buildings and the way in which their buildings 'communicate' with building users and those passing by. This is an important and necessary part of understanding architecture, although it is unusual to find other contributors to the AEC sector adopting this interpretation of the word 'communication'.

This chapter is concerned with the way in which project participants communicate and the media and various languages employed to achieve mutual understanding. The word 'communication' is used throughout this book in the way that it is used in the majority of the communication literature, i.e. communication is the sharing of meaning to reach a mutual understanding and (usually) to gain a response (Emmitt and Gorse, 2007). This involves some form of interaction between sender and receiver of the message. It is important to note that the word is not used in the way it is in some of the management literature, in which the verb 'communicate' is used as a synonym for 'transferring' information to another party.

The creation of meaning between two or more people at its most basic level is an intention to have one's intentions recognised (Sperber and Wilson, 1986). Informing someone by any action that information is to be disclosed is considered to be an act of communication. Sperber and Wilson (1986) suggest that when people communicate they intend to alter the cognitive environment of the person whom they are addressing. As a result, it is expected that the receiver's thought process will be altered (unless they choose to ignore the message).

The process of communication

Early communication models (e.g. Shannon and Weaver, 1949) used a mechanistic Sender–Message–Channel–Receiver model, in which the sender sends a message via a communication channel to a receiver. The message may be distorted by noise (interference) as it is sent. Once received the receiver will decode and interpret the message, with no certainty that the intention was fully understood. Although this implies a linear process, the reality is that communication is an iterative process, with messages constantly being sent and received in an environment full of distractions (noise). Interaction is necessary to check that the message received is the same as that sent, which can be achieved quite quickly in conversation, but which can take longer when exchanging drawings, figures and text.

Simplistic models may be useful in helping to understand the fundamentals of the communication process, but they may also be misleading. Communication is concerned with understanding. Rogers (1986) proposed a shared perception model in which each person is a participant, rather than a sender or a receiver, and this appears to be more relevant to a multidisciplinary project environment in which individuals are communicating to realise the project goal(s). Participants will operate selective exposure to messages, choosing to respond to or ignore messages depending upon their circumstances and needs at a given time, a characteristic termed selective perception (Hassinger, 1959).

For understanding to take place most theorists claim that a background of shared social reality needs to exist. To engage in meaningful communication we need to build on information and develop a context supported by cues and clues. These guide us to use subsets of knowledge and help us to link information together. Clues used come in many different guises. For example, Brownell *et al.* (1997) suggest that the appropriate mode of referring to something or someone in conversation depends on what common ground a speaker and addressee share. It follows that disciplinary groups will refer to different cues and clues from a multidisciplinary group, as their level of common ground will differ. Thus we need to adjust how we communicate depending on the context.

It is the lack of common areas of understanding and a failure to develop a shared understanding that lead to ineffective communication. Given the multidisciplinary nature of AEC projects it will be necessary for the participants to devote time and effort to explore common ground and develop a shared understanding at the start of new relationships. The speed at which and manner in which this is achieved at the start of projects, and at stages in projects when there are major changes in personnel, will influence the effectiveness of communications and the effectiveness of the TPO. Team-building workshops and introductory project meetings are common approaches to stimulating interaction. Failure to explore common ground is likely to lead to ineffective communication, problems with coordination of information and errors, all of which tend not to be evident until a problem arises.

When there is a lack of congruent understanding the speaker has to provide an infrastructure of contextual information on to which the new information can be built, developed and hence understood. During interaction the speaker has to access and build a framework of the other's knowledge and influence the way the recipient draws together the subsets of knowledge to make communication work. Using clues sent from the receiver of the message the sender may make assumptions of the knowledge a person possesses. This initial and often tentative interaction can be used to check whether early assumptions are correct at a critical stage in the development of a relationship. Offering information, opinions and beliefs and asking questions helps to develop an understanding of the other person's knowledge and beliefs, as does examining and analysing the verbal and non-verbal information presented (Bales 1950, 1970). Failure to exchange information and ask questions will decrease a person's ability to use the other's knowledge and thus considerably hinders effective communication (Brownell *et al.* 1997). The importance of cues and clues that are embedded in the context, body language and emotion play a key part in the development of understanding and hence the relevance of exchanges.

Reluctant communicators

As individuals we operate selective perception and hence choose to accept or reject the messages we are constantly receiving. We also exhibit a variety of traits when it comes to communicating, with some of us more forthcoming than others. Our enthusiasm for communicating is linked to our personality and the situation; indeed, participation is related to an individual's willingness to speak. Take, for example, a project manager who likes communicating when things are going well, but who becomes apprehensive and withdraws from interaction in a crisis, just when it is most needed.

People who avoid communication, either in certain situations or as a personal trait, are termed reluctant communicators (Wadleigh, 1997). Shyness may occur because of communication discomfort, fear, inhibition and awkwardness (McCroskey, 1997). So, although we may be happy to communicate with our peers and colleagues within the organisation, we may well exhibit some reluctance to converse in a project environment if we feel uncomfortable. This could simply be the fact that we attend a meeting and are unfamiliar with the other attendees, or it may be that we feel threatened by a crisis or dispute. In group situations reluctant communicators talk less than others and are unlikely to initiate communication. Highly apprehensive individuals will also have a tendency to avoid meetings and workshops if they can.

Interdisciplinary project communication

Communication in any group has social and task dimensions. Task roles are those that determine the selection and definition of common goals, and the working towards solutions to those goals, whereas socio-emotional roles focus on the development and maintenance of personal relationships among group members. Open and supportive communication is conducive to building trust and facilitating interaction between project members. Unfortunately, when communication between team members is most needed, during times of uncertainty and crisis, relationships often break down through the development of defensive behaviour and hence ineffective communication. Open exchanges of information and sharing task responsibilities are essential for effective working relationships. Interaction that builds and maintains the fragile professional relationships within a project that are necessary to accomplish tasks is fundamental to project success.

Interdisciplinary communication

Interdisciplinary communication can be explained as a series of interactions between a group of participants (senders and receivers) using a web of communication channels and an assortment of media and tools. Communication involves some form of interaction between the sender and the receiver of the message through synchronous and/or asynchronous communication:

* 'Synchronous communication' is the term used when individuals and groups communicate at the same time through face-to-face dialogue and interaction in meetings, telephone conversations and video/online conferencing. Synchronous interaction is essential for addressing contentious issues, problem solving, conflict resolution, exploring values, developing trust and building relationships.

- 'Asynchronous communication' is the term used when parties do not communicate at the same time, for example by email, sms (text messaging) and intranets and by post and facsimile. Messages are sent and (hopefully) responded to sometime later. Asynchronous communication is essential for transferring a large amount of information to the receiver(s) when an instant response is not required or is not possible. Receivers of the message will be able to respond (if they feel they need to) at a later date once they have had time to assimilate the information and consider their response.

In a project environment it is common to use both synchronous and asynchronous communication depending on the task, the preferences of the participants and the stage of the project. For example, face-to-face communication through dialogues and meetings will be used extensively in the early design phases when exploring possibilities and when consensus is necessary for the design to proceed. Later in the project, for example during the production information stage, the emphasis shifts to producing information and information exchange, so the emphasis is more on asynchronous communication. The challenge appears to be associated with identifying how members of the TPO prefer to communicate within design and construction projects. According to Otter and Emmitt (2007) effective design teams use a balanced mix of synchronous and asynchronous communication. Balancing the use of available communication media may need to be carried out more than once during the project, depending on its phase. This requires consideration of the usefulness of high or low levels of interaction, the feedback required to stimulate progress and the risks of miscommunication.

Communication levels

One way of helping to identify communicators and communication routes is to analyse the TPO in terms of communication levels and communication channels (discussed later). Communication can be separated into five levels (Emmitt and Gorse, 2003), from the most private (intrapersonal) to the most public (mass communication):

- *Intrapersonal (private and hidden).* Intrapersonal communication is an internal communication process (cognition) that allows individuals to process information. Only one person is involved in this thought process, which is usually private and hidden from others. This is sometimes referred to as the 'black box' because our thoughts and reasoning are not accessible to others, nor for that matter are our hidden personal and organisational agendas.
- *Interpersonal (intimate).* Interpersonal communication is between two people (a dialogue), which allows individuals to establish, maintain and develop relationships. Conversation is an intimate process

during which common ground can be established quickly through the common understanding of terminology and language. It is also through interpersonal communication that we tend to make judgments regarding the trustworthiness of others, a point taken up in Chapter 3.

- *Group (familiar)*. Group communication occurs in small groups or teams, which may be disciplinary or multidisciplinary in their constitution. These small groupings of individuals are usually able to develop effective communication quite quickly as they work towards a common objective. Groups are usually, but not exclusively, based in the same organisation and so they are familiar with the organisational culture and protocols. An exception would be a creative cluster, a grouping of individuals from different disciplines and organisations assembled to engage in creative problem solving (see Chapter 5 for a fuller description).
- *Multigroup (unfamiliar)*. Multigroup communication occurs within social systems such as organisations and TPOs. With the exception of small organisations and TPOs, these social systems function because of the collective efforts of interdependent groups and teams, working to achieve a common goal, which is also termed 'intergroup' communication. The manner in which these groups and teams interact and their familiarity with the other organisational cultures and protocols will influence the effectiveness of their communications. In large organisations projects may span departments and sub-groupings of staff. In TPOs small groups and teams communicate with other, often unfamiliar, groupings, only getting to know how they like to work as the project develops.
- *Mass (public)*. Mass communication involves sending a message to large audiences. This may involve advertising through the mass media (professional journals, newspapers, television, radio, internet) or the dissemination of important information, for example changes to health and safety legislation, to the entire TPO via the project intranet.

All five levels serve different functions and all are equally important in achieving effective communication and in limiting the amount of ineffective communication. In the context of this book the most pertinent communication level is multigroup communication or interorganisational communication. In multidisciplinary teams, members come from different organisations, which have different organisational cultures and which also use a variety of information systems. Individuals also have different levels of understanding, opinions, skills and rates of adoption of communication technologies, as well as preferences for specific means of communication (e.g. Tuckman, 1965; Gorse, 2002).

Communication channels

Another means of analysing communication within the TPO is to look at formal and informal channels of communication:

- Formal communication channels are those set up and reinforced by the procurement route and legal contracts. Individuals are expected to follow certain protocols and use communication channels that have been established for the project. These may be very similar to, or subtly different from, the protocols and communication channels used within each contributing organisation. This means that individuals may have to adjust how they communicate across different projects. Formal communication routes can be designed before the commencement of the project and mapped during the project to ensure that they are appropriate for the project context and project stage.
- Informal communication routes are those that develop around the formal channels and are established by communicators to help them do their job more effectively. Research has shown that in times of crisis project participants tend to favour informal communication routes to try and resolve the problem, conscious that formal communications are recorded and may be used as evidence if the problem turns into a dispute. The challenge for managers is that informal communication channels by their very nature are difficult to track and individuals tend to circumnavigate formal procedures, such as those set down as part of a quality assurance (QA) plan.

Communication networks

Communication channels form part of the communication networks that exist within organisations and within the TPO. It is common practice to use network analysis techniques to determine the communication structure within a social system, from which communication networks may be represented graphically. This is expressed by linking nodes (communicators) with other nodes, with the most frequent communicators represented by larger nodes and the less frequent by the smallest node. Analysis of the frequency of communication between individuals also allows the most active communication channels to be identified, usually represented as lines of varying thickness to represent frequency of communication. These techniques can be useful in helping to map communication routes and hence manage communications more efficiently. The disadvantage of network analysis is that it does not identify the quality of the communication, merely the quantity. Another challenge with network analysis is that the results tend to represent a network at a fixed point in time, the time that the analysis was conducted, and do not address the change in the network over time (Rogers and Kincaid, 1981). According to Rogers and Kinkaid communication networks are so fleeting that they cannot be accurately mapped. This is especially true of AEC projects, in which the social system is highly dynamic and fragile, thus network analysis would need to be conducted very frequently to capture the extent of the connections. To some extent ICTs can provide a good picture of communication activity within a project environment, but the problem is that ICTs do not capture the informal communication that happens around the edge of the formal communication channels. Therefore our network maps can provide a reasonable indication of project communications, but cannot show the full extent of communication within the project network.

Another way of looking at communication networks in AEC projects is to categorise the contributors to the TPO by their involvement with the project (Emmitt and Gorse, 2003):

- *Formal (contractual)*. The formal network is constituted through legally binding contracts. Organisations and individuals are clearly identified and their project roles defined by contractual agreements. These operate on a number of levels. For example, deciding to use a particular type of curtain walling will establish communication routes with the manufacturer and the specialist sub-contractor(s). Eventually, this will be formalized by the signing of a contract.
- *Statutory*. Various external contributors to the project are represented by individuals working for statutory authorities, such as the town planning officer and the building control officer. These will be determined by the physical location of the construction site, as well as the type and complexity of the project. These contributors have no formal contract with the project, but will nevertheless influence the project through the ways in which they interact. For example, the town planning officer my be supportive or resistant to a proposal based on the council's town planning policy, and this can lead to harmonious or antagonistic relationships.
- *Informal*. The informal network will comprise local interest groups and user groups, who will participate in the project through cooperation or through protest. By their very nature these groups can be very difficult to define and communication can be intermittent, unexpected and hence challenging to manage. Good managers take a proactive approach and try to include consultation events that allow the various contributors to interact as part of the project process.

Open and closed communication

The manner in which project participants choose to communicate also needs to be considered. At different times in the life of the project and in different situations it is to be expected that individuals will choose to be open or closed (defensive) in the way that they communicate with others. Communication among familiar colleagues and in small groups and teams that have been established for a reasonable amount of time tends to be relatively open, with individuals engaging in candid and frank intercourse. When individuals venture out of their familiar environment they tend to be more guarded and less candid. When individuals perceive a threat they tend to resort to defensive communication, in which the information disclosed cannot be used against them.

Language

Communication is essentially a social activity, the sharing of information and the sharing of experiences, which is dependent upon the communicators understanding the rules of communication. Speech, writing and drawing are obvious modes of communication, but so too is body language, which can convey more subtle, and rarely recorded, understanding. Language is central to all social activity and involves abstract notions, actions and events removed in time and space, with subtle shades of meaning and logical distinctions that depend on people sharing a complex and symbolic representational system (Potter and Wetherell, 1987).

The words that we use in spoken communication, and how we choose to deploy them using tone and intonation, can make a significant impact on how individuals are perceived by others, and can affect the level of understanding. Mehrabian (2007) highlighted the importance of meaning as distinct from words. According to Mehrabian's communication model, less than ten per cent of meaning is in the words spoken; the majority of meaning is conveyed in style, expression, tone, facial expression and body language. The way in which words are said (paralinguistics) accounts for almost forty per cent of meaning and the remaining meaning is conveyed in facial expression. This helps to illustrate the importance of visual signs when trying to convey or interpret meaning. It also helps to emphasise the importance of face-to-face communication in a physical setting. In situations in which visual channels do not exist, for example telephone communications and written text (e.g. emails), as a general rule there is a greater chance of confused understanding and inferred meaning. Videoconferencing provides an alternative to face-to-face meetings, although because of the intermittent transfer of images there may be some loss of non-verbal signals.

In multidisciplinary temporary project organisations it is inevitable that participants use language in a variety of ways and hence speak a different language from their fellow participants. This is sometimes seen as a weakness of multidisciplinary working, although it is merely a manifestation of the different knowledge and backgrounds brought together to realise the project. Managed proactively the differences in language use can be a considerable strength of the project.

Disciplinary languages

Professionals and tradespeople have developed specialist languages, using words and terminology that are specific to their discipline; essentially a unique vocabulary. This enables members of the same discipline to communicate specific facts and ideas quickly and efficiently. For example, architects are often accused of using 'archispeak' by those outside the discipline.

Words can have very different meanings within specific professions and trades, although the communicators rarely define the words that they are using. Stretton (1981) acknowledges that these aspects constitute a major barrier to effective communication and urges participants to consider the receiver of the message and communicate in a language that is perceived to be acceptable to them. This involves explaining and developing an under-. standing of associated issues before the main event is discussed. Participants need to acknowledge that differences will exist within an interdisciplinary project environment and consciously adjust their communication to appeal to a broader, multidisciplinary audience. Confusion can arise when workers from different parts of the UK use different words for the same thing, such as a building component, and regional dialects offer the potential for further misunderstanding. Fortunately, during conversations there is usually an opportunity to ask for clarification if the words used are unfamiliar to those taking part in the exchange. With written and graphical communication the opportunity to ask questions is usually less immediate and the potential for misunderstanding is greater.

'International' projects

Migration of construction workers within Europe and across the globe has resulted in many construction sites with a multinational workforce. Language will take on even more significance in 'international projects', a term used to describe a TPO made up of organisations and/or participants from more than one country, thus creating a multicultural TPO. The potential for misunderstanding is never too far from the surface in such environments, even when everyone's intention is to do a good job.

To enable the project to function effectively one language must be used for all project communications. English is often the language of choice, simply because many cultures have English as a second language. This means that the contract, formal meetings and all project information must be in English for the project to be managed effectively. Thus many of the participants may be communicating in a language other than their native one (their second or third language), in which words may carry different meanings. Similarly, the significance of words and how they are uttered can mean very different things to other participants. Given that the possibility of misunderstanding is likely to occur, it is necessary for all participants to continually monitor how they express themselves and how messages are reinforced with other media, such as drawings.

It is at the boundary conditions, the interfaces, that individuals may well have to use different language to express themselves. On a day-to-day basis it is common for the participants to interact using their native tongue and on construction sites it is not uncommon for gangs of workers to have one spokesperson, an interpreter, who can communicate with the other gangs and the managers.

Media

The effectiveness of project communication appears to be influenced by a number of inter-related factors, such as an individual's preference for using specific (codified) language, his or her preference for favouring the use of some communication media over others (e.g. drawings over written specifications) and the ease of access to user-friendly communication tools such as project web technologies. Given the large number of individuals involved and their various preferences this could, if not managed proactively, give rise to ineffective communication.

To ensure that a message achieves its desired effect it is essential that the method used to transfer information supports the communication process. Choice of communication media can play an important role in the ability of project participants to understand one another and hence work effectively. For example, plans and detailed drawings may be the design team's preferred choice of media, but the information codified within the drawings may be incomprehensible to clients who have no experience of AEC projects. In such a situation a physical or virtual model might be a better choice of media to explain key concepts and spatial arrangements. Similarly, individuals from different disciplines may have preferences for text and figures over drawings or vice versa. A survey of media use by construction professionals (Gorse, 2002) helped to illustrate some subtle differences between professionals' preferences for certain types of media to communicate their ideas to others. For example, some professionals preferred using the telephone to email, others preferred face-to-face meetings to telephone discussions. Combined, the professionals surveyed by Gorse (2002) reported a preference for face-to-face communication, followed by letters and drawings, over all other means, such as telephone and email. The sample also claimed that informal meetings were the most effective means of communicating during the construction phase of a project. The important point to make is that one medium may not necessarily be better than another *per se*, but may be more effective at communicating a message for a given context and a given audience.

Communicators should consider a number of fundamental issues before choosing from the available media (summarised below). Of particular concern is the reason for the message: is it to exchange ideas or is it to convey instructions and/or information? The answer to this question will help in identifying the recipients of the message and the most appropriate, or effective, media to facilitate communications. Another issue relates to the formality, or informality, of the communication, and the choice of media best suited to convey and (if required) create a record of the communication. The choice of media also needs to consider the users and their ability to use the media in the workplace, be it an office environment or a construction site. Time is another determinant. In well-planned projects there is time to think about the best medium to use and time to prepare communications, such as drawings and written specifications. However, there will be occasions when an unexpected event demands a rapid response, in which case there may be insufficient time to prepare a drawing, but a very quick sketch may be enough to convey appropriate instruction to other members of the TPO.

Oral communication

Oral communication skills are essential for explaining ideas, exploring problems, coordinating work packages, resolving disagreements and getting things done. The ability to converse with others also has a social function in terms of developing effective working relationships. One of the main benefits of conversing with others is the ability to ask questions and explore uncertainty, which can be achieved quickly and can avoid the need for lengthy description by, for example, email. The main means of communicating orally are:

- telephone
- video/web conferencing
- dialogues
- meetings and reviews
- workshops.

Written communication

The ability to write clearly and effectively for a variety of complementary purposes is an essential requirement. The level of formality may vary, for example between an email to a close work colleague and a formal instruction by email to a contractor to change the design, but the purpose of the written communication is the same. Some forms of written communication, such as the written specification, must be written in a specific manner so that all users of the specification clearly understand the contents of the document (see Emmitt and Yeomans, 2008). Written communication includes:

- emails
- letters
- reports and meeting minutes
- instructions and variation orders
- specifications
- schedules
- contracts.

Graphical communication

The majority of design ideas are communicated through graphics. Drawings are one of the most familiar and effective means of communicating information within the TPO, although as with other forms of codified information the interpretation of the information can vary between users unless clear and accepted drawing protocols are followed. Standard conventions allow all users to quickly understand what is required. The ability to draw and to read drawings varies considerably within the TPO. Architects are able to communicate very effectively with conceptual sketches, but more information is required if the same meaning is to be conveyed to engineers or surveyors. Clients unfamiliar with construction may be incapable of reading drawings and so virtual and physical models may be needed to communicate design intent and represent the proposed form of the building. The main graphical tools used are:

- sketches
- two-dimensional and three-dimensional drawings
- perspective drawings
- virtual models.

Physical representation

Physical scale models are an effective means of communicating space and form in three dimensions. They are used to help develop designs and also to communicate with individuals who are less able to understand technical drawings, such as clients and members of the public. Sample panels are used on the construction site to check and approve samples of materials and the quality of work before work commences.

Electric theatre

As computer hardware and software has developed it has become common to communicate using electronic communication tools. Email has become an essential tool for disseminating and requesting information asynchronously, and also for contacting people who are not easy to contact by telephone. Project intranets and extranets have developed into sophisticated tools to help all participants manage the vast amount of information generated for AEC projects. Despite the rapid development of electronic communication systems challenges still remain in terms of interoperability of software and bandwidth (technical challenges), as well as the manner in which people use the tools (social challenges). The technical challenges will, in time, be resolved, but the social challenges are likely to remain a constant factor.

Care is required when planning and implementing ICTs to ensure that the system(s) meet the needs of the many and varied users. Unfortunately, the use of new technologies alone does not necessarily ensure better performance, especially when there is a mismatch between the system and the preferences of the users. Implementation of new IT tools and software will require investment in training and time for users to become familiar and competent with the new technologies. In an organisational setting it is usually possible to introduce new technologies with little disruption to familiar working practices, but this is not often the case in a project context. Individuals may find themselves working on a project with unfamiliar software applications, and this can have implications for the efficacy of interaction.

As mentioned earlier, it is the interpersonal, often informal communication that forms the glue between individuals and organisations, in addition to the tools to facilitate communication, which is important. The implication in this statement is that those designing the communications infrastructure need to map the communication networks before assessing the various tools available and deciding on those tools most likely to benefit the project. Research into IT and its effect on productivity has shown that new technologies do not always result in improved productivity (Brynjolfsson, 1993), a phenomenon known as the 'IT productivity paradox'. This problem is further complicated by another well-known problem, the rivalry of IT tools, or the lack of interoperability between software applications. This can be a serious obstacle to communication in a project environment when different organisations use different technologies. One way of overcoming this conflict is for all project contributors to use a common ICT platform, a project website.

Project websites

Project websites (extranets) can provide an important vehicle for project contributors to share information and develop their work concurrently. Members of the design team are able to share information files and access current work of other designers; to be effective the project website needs to be designed to suit the needs of the designers and managers using it and once it is implemented training will be required to allow all project participants to use the system effectively and efficiently. Failure to follow accepted protocols may result in problems with coordination activities. The quicker that information can be accurately produced, exchanged and understood by those dealing with it, the better. Much attention has been given to the transfer of information within project environments, as the timely exchange of information is a key factor for effective coordination and hence control of project information. A considerable amount of information can be processed at remote locations, for example in the office, where all the tools, equipment and information are to hand, and information produced can be easily sent by email or web-based technology. Increased use of intra-webs and project-extranets has made a significant impact on the way in which information is exchanged between project actors. It would be prudent, however, to remember that people produce information and we are all prone to making mistakes, especially when working under pressure. Subsequently some form of interaction between the sender and the receiver of the information is required to coordinate information and resolve discrepancies before it can be approved for use by others.

Electronic conferencing

Telephone conferencing and videoconferencing can be a quick and effective means of allowing several people from different geographical locations to communicate without having to travel to a meeting venue. Telephone conferencing is easy to arrange, but the opportunity to see facial expressions and body language is not possible and so the richness of communication is compromised compared with face-to-face interaction. Videoconferencing mitigates this problem somewhat because it is possible to see facial expressions and some body language, depending on the position of the communicators in relation to the cameras.

Collective communication framework

Otter and Emmitt (2007) found that a collective framework for project communication and collaboration using electronic tools was missing. There was also evidence of a lack of understanding by the users of the proper use of the tools, a lack of training and poor management competences to stimulate proper use. Furthermore, the rivalry of tools tended to hinder, rather than improve, effectiveness of communication within the TPO. It would appear that a bottom-up approach to the management of communication is required to improve effectiveness (Otter and Emmitt, 2007). Developing a common understanding of effective communication within the TPO and using the most appropriate means for the purpose is fundamental to project performance.

Practical challenges

Communication is dependent on the willingness of all participants to act and react and to listen and share, as well as develop their skills for using communication effectively (Forsyth, 2006). It follows that project communication is likely to be most effective when all members contribute using the available communication media in the same way, and as agreed to at the start of the project, i.e. they follow project communication protocols. Experience tends to suggest that this is an ideal, rather than a reflection of reality. Both managers and project contributors need to understand and agree to systematic communications based on collectively agreed rules for a specific project context; this should bring about both individual and collective benefits for the contributors (Otter and Emmitt, 2007).

Leading and stimulating effective communication is a challenging task for a number of reasons. First, the number of electronic tools is increasing and therefore both users and managers need to develop specific skills for their collective use (Otter, 2005). Second, there are differences between the participant organisation's use of electronic information systems; combined with the variety of communication practices this may create problems with compatibility. Third, there are differences in opinions and understanding on an individual level, including differences in the use of specific ICTs. Combined with the lack of a collective framework for meaning (Mulder, 2004) these factors can lead to misunderstanding within the TPO.

Research has shown that the most important skills for a project manager are the ability to communicate and listen, closely followed by the ability to negotiate, influence and persuade. Good practice in TPOs is characterised by open exchanges and transparent communication, enabling responsibilities and ownership of tasks to be negotiated and shared. The challenge for managers and participants alike is to identify and hence tackle ineffective communication while simultaneously promoting effective communication – no easy task in a dynamic project environment. The AEC sector is notorious for its litigious and fragmented culture, which cultivates closed and defensive communication behaviour. Although this is not applicable to all projects, it needs to be considered when suggesting ways of improving communication.

Communication breakdown can occur for a variety of reasons. Some of the most common errors relate to misunderstanding, the failure to communicate at an appropriate time (or failure to communicate at all), the wrong people communicating and the failure to ask questions (Emmitt and Gorse, 2007). Time pressures are a constant threat to effective communication; everyone appears to want answers and information immediately, and this can lead people to issue information before it has been checked and/or to say something without first checking that the message is accurate. To a certain extent good managerial frameworks and the appointment of individuals who understand one another will help to limit communication breakdown, but it will happen to lesser or greater extents regardless of how organised the TPO is.

The majority of books that deal with various aspects of communication, such as report writing or production of information, tend to offer some generic golden rules as an aide-memoire to those trying to express their ideas to others. Not surprisingly, these generic rules tend to apply to the majority of exchanges and this is the approach used here. The golden rules for achieving effective and efficient communication are:

- *Accuracy*. Accurate use of language, terminology and conventions will help to avoid confusion and will reduce the likelihood of errors resulting from misunderstanding.
- *Avoidance of repetition*. Repetition is wasteful and can lead to confusion because the 'same' message communicated through different media and by different participants will be interpreted slightly differently.
- *Checking*. Failure to check information and communications for errors and compliance with appropriate standards, codes and project-specific requirements is a common cause of rework, which is wasteful.
- *Clarity and brevity*. The skill is to convey only that which has relevance and hence value to the intended receiver. This calls for an appreciation of the receiver's requirements.

- *Consistency*. Given the large number of disciplines contributing to a TPO it is crucial that the use of words and language is reassuringly consistent across the entire project.
- *Timeliness*. All communications should be timely, which means that consideration should be given to the flow of information during the project and the timing of meetings and workshops to aid coordination and collaboration.
- *Question*. Question anything that appears to be wrong, or when the meaning is not clear.

End of chapter exercises

- Identify a person whom you perceive to be a good communicator and list up to five characteristics that you feel make them a good communicator. Now identify a person who you feel is a poor communicator and list up to five reasons for your choice. Compare and list the differences between the good and the poor communicator. Now try and position yourself in relation to the good and the poor communicator: what are your strengths and weaknesses?
- You are participating in a meeting to discuss design development drawings with a wide range of stakeholders. During the meeting it becomes evident that the client's representative and the user's representative are struggling to read the drawings, although they have not said anything. What do you do?
- We should be concerned with contacts, not contracts. Do you agree or disagree with this statement? Please justify your answer.
- Procurement routes and contracts help to establish lines of communications, interfaces and relationships. How would a project manager ensure that the interfaces are working as planned during the life of the project?

Further reading

Additional material relating to communication in AEC can be found in *Communication in Construction: theory and practice* (Dainty *et al.*, 2006), *Construction Communication* (Emmitt and Gorse, 2003) and *Communication in Construction Teams* (Emmitt and Gorse, 2007). For additional information relating to ICTs see *Construction Collaboration Technologies: the extranet evolution* by Paul Wilkinson (2005).

3   Trust

It is through interaction and communication that we are able to establish the trustworthiness of our fellow project participants. Trust, or the lack of it, underpins both personal and business relationships and according to Brenkert (1998) allows organisations to reduce transaction costs, share sensitive information and engage in joint projects. In a work environment relationships are governed by money and power, or more specifically the distribution of resources. This is common to organisations and TPOs. Tensions manifest as individuals seek to control resources and exercise power to their, or their organisations', mutual advantage. Although it might be convenient to ignore such struggles, and hence distort our understanding of the situation, it should be recognised that this creates tension between individuals and groups, which has a habit of surfacing in times of crisis.

In the AEC sector trust has become a topical issue as organisations have started to implement ways of working that try to confront the sector's adversarial culture. Similarly, the focus on delivering value and the attention to corporate and personal values also helps to highlight issues concerning trust and transparency. A collective inability to trust other project contributors is considered to be a major challenge facing the AEC sector (e.g. Bennett and Jayes, 1995; Latham, 1993), although this is also true of others. Lack of trust has been found to be a problem in the service and manufacturing sectors (Morris, 1995) and also within organisations (Karl, 2000). With organisations using flexible working practices and adopting project-orientated approaches the issue of trust between temporary project groupings has started to attract more attention from researchers and practitioners. This has also brought about greater attention to the values and attitudes held by individuals participating in temporary work groups and attention to developing trust within the TPO.

Cooperative arrangements such as project partnering and strategic partnering, together with the ideal of integrated teamworking, are founded on trust, or more specifically trust between business organisations. At the start of these relationships it is reasonable to assume that trust does not exist (unless individuals have successfully worked together previously), but that trust will develop and be tested as individuals and organisations start to interact. Trust also exists in competitive arrangements at various levels. For example, it would not be sensible to include an organisation on a shortlist of tenderers if it could not be trusted to perform. In temporary project relationships we have to learn to trust the people that we work with and they have to trust us, which is often much easier to state than it is to achieve in practice.

The opposite of trust is (managerial) control. Business relationships are often based on low levels of trust and a certain amount of distrust, which is why we have legally binding contracts that set out the rules of engagement. There is nothing wrong with this type of relationship *per se*; indeed, one could argue that everyone knows where they stand when there is little or no trust between the parties to the same contract. Rules of engagement are known at the start of the contract and there are procedures to mitigate the lack of trust. A common argument is that projects based on low levels of trust are expensive in terms of the procedures adopted and the overall value delivered to clients, although it is possible to achieve successful projects with no or very little trust. Trying to engineer trust within the temporary project team is expensive and takes considerable effort, and of course there is no guarantee that things will go to plan (which is why there are partnering contracts).

At both ends of the scale, total trust or total distrust, we know where we stand, although such absolutes are quite rare in practice. Somewhere in the middle range of the trust scale can be problematic and challenging to work with unless we establish some clear parameters. These parameters can be established only through interaction and communication, as we start to work with other project participants in co-located, virtual and global teams. What makes this area particularly fascinating in terms of interdisciplinary management is that projects comprise a dynamic mix of relationships based on varying degrees of trust, and these boundaries or parameters tend to shift over the course of a project as project participants interact, test one another and react to their individual and collective experiences. The dynamic within and between groups will have a bearing on how trust develops and trust may be one of the characteristics that influences group and intergroup performance (Serva *et al.*, 2005).

To survive and prosper in the workplace we have to manage conflicting demands and priorities. No two individuals' needs converge as much as we like to think and we often live out a sort of pretence in the workplace to give our relationships credibility, i.e. we sometimes stretch the truth to help maintain our relationships with others. This stretching of the truth may involve omission, exaggeration and deliberate ambiguity. Judgment about trustworthiness is not just about whether someone tells the (absolute) truth all of the time; it is also concerned with sincerity and consistency of actions. Context is a crucial determinant. For example, we might admit to a close work colleague that our performance is hindered because we had too much to drink the night before, but we are unlikely to disclose this to our line manager. Similarly, we might be candid in the office about the poor performance of the project, but it may not always be diplomatic to tell the absolute truth in a project progress meeting.

Fundamentals

Despite a large literature on the subject of trust the term appears to be open to considerable variation in interpretation and application. It is, however, quite common to define trust as a firm belief in the reliability of a person or a thing. Or put another way, trust is an attitude held by the truster toward the trustee (Spector and Jones, 2004). The perceived level of trustworthiness will be based in the form of beliefs about the trustee's integrity, ability and benevolence (Mayer *et al.*, 1995; Serva *et al.*, 2005). The perceived level of trustworthiness will come from interaction and communication with others and experience of the way that others act and behave. For example, we trust what someone tells us if we feel that the message comes from a reliable source; we trust someone to carry out a task because they are perceived to be reliable; we trust someone because they have always acted with integrity. The implication here is that we need to have some experience (either direct or indirect) of a person or an organisation before we can start to trust them. Indeed, it would be unwise to trust people we do not know because they may prove not to be motivated by good intentions; and we are unlikely to know what their intentions are until we have worked with them for a period of time. This is why it is necessary to seek personal references and testimonials before appointing new people to an organisation or selecting organisations to work on a project.

Because trust is contextually derived it is not easy to conceptualise trust in absolute terms (Atkinson and Butcher, 2003). Trust is largely about our confidence to trust others' commitment to shared project goals and values. Trust is concerned with uncertainty, vulnerability and risk (Meyerson *et al.*, 1996). We are more likely to trust someone from the same discipline than someone from another profession simply because we understand their culture and trust them to act in a certain (familiar) way. Hopefully, we can trust what a fellow project participant tells us and trust them to complete their task on time and to the agreed quality. By placing our trust in others we are vulnerable. For example, it is not uncommon to hear people say something like 'they will only let me down once' when working with people they do not know too well; essentially a recognition of their vulnerability and the fact that trust has to be earned.

Limits and boundaries

Trust is not unlimited; we trust some people more than others and with different aspects of work and business. We do not trust an organisation *per se*; rather we trust the individuals working in the organisations with whom we have contact. When actors fail to live up to expectations trust starts to erode, and it is very difficult to regain the same degree of trust once it has been damaged. Once trust has gone there are two choices, either to fall back on systems and procedures for control (such as legally binding contracts), or cease the relationship (which is likely to be disruptive in the short term). This means that some tough decisions will have to be made once our confidence in the ability to trust someone has evaporated.

The development, learning, testing and reaffirmation of trust will usually require personal contact, although Jarvenpaa and Leidner (1999) have demonstrated that this can be achieved in global teams and with no interpersonal contact. Within an organisation there is regular interaction between staff and the degree and levels of trust are usually well understood; indeed, many professional offices rely on trust and mutual respect in preference to rules and regulations to achieve their objectives. By contrast, management by trust is not easy to achieve in a project context. With people interacting only occasionally and holding different organisational values and objectives, the development of trust is far more challenging because participants often have very limited opportunities to get to know others well enough to develop an adequate level of trust. This is why we have contracts.

- Within organisations trust will be developed over time as relatively stable relationships are developed within small groups and sub-groups of employees. Experience of how others respond in a particular organisational environment is important in developing trust.
- Within the project environment relationships are less stable and participants work in different organisations on a daily basis. Compared with the members of a business organisation the project participants interact less frequently, perhaps once per week, and therefore it may be difficult for individuals to develop trust based on experience. Thus individuals will develop a different kind of trust, swift trust, which is described in more detail below.

It is important not to confuse trust and distrust with liking or disliking someone. Nor should we confuse trust and distrust with the ability to agree or disagree with our fellow project participants. It is entirely probable that during our working lives we will find ourselves in project relationships in which, for whatever reason, we take a dislike to one or more of the participants with whom we have little option but to interact, but we may still trust them to do their job and act with integrity. Conversely, we may like interacting with other team members but not trust them to deliver their work on time. It is also possible to trust someone but not agree with their views all of the time; indeed, it may be easier to engage in a critical and constructive discussion with people that we trust than with those we are suspicious of.

Initial trust

It is the start of new relationships in organisations (Spector and Jones, 2004) and TPOs that is the most critical time for the development of interpersonal trust. Initial trust among employees is determined by an individual's trusting stance and on a number of variables associated with the context of the situation (summarised below). Keller (2001) found that initial trust levels tend to continue, in either a negative or a positive direction, as a relationship develops. Thus the formation of initial trust is a significant factor in the development of interpersonal trust.

Trusting stance

Trusting stance is a conscious choice to trust others until they prove to be untrustworthy (McKnight *et al.*, 1998). According to Spector and Jones (2004) the findings of empirical research have not established whether or not trusting stance leads to the development of interpersonal trust, although in their study of 127 professionals working in the north-east United States using simulated role play they found that trusting stance was positively related to initial trust levels in the workplace, as posited by McKnight *et al.* (1998). It would be reasonable to assume that the trusting stance of the individual members of a new TPO will determine the initial trust levels.

Category-based trust

The type of trust an organisation places on employees (insiders) is termed category-based trust (Kramer, 1999). From this, Spector and Jones (2004) speculated that internal employees might be given higher levels of trust by an organisation than outsiders. This is because the insiders are assumed to work under the organisation's rules and norms and the outsiders are not. However, they found that the initial trust level was not affected by whether the individual was internal or external to the organisation. Although more research is required into this aspect of trust, it could be a factor in interdisciplinary project organisations.

Role-based trust

Role-based trust is influenced by knowledge of a person's occupation and their specific role within an organisation (Kramer, 1999). Individuals tend to exhibit higher levels of trust for their superiors than their peers because there tends to be more dependency on them than on their peers, although Spector and Jones (2004) did not find this to be the case in their study.

Gender

The role of gender also needs to be considered in the development of initial trust. Some studies have found trust levels to be higher within the same gender (men will trust men more than women; women will trust women more than men) whereas others have not reported any differences associated with gender. Spector and Jones (2004) found that men demonstrated a higher initial trust level for other men than they did for women. They found no differences between the women who took part in their study. This is an area in which more research is needed; however, it is worth remembering that individuals might be less trusting of the opposite gender (Spector and Jones, 2004).

Trust within temporary project organisations

Baden Hellard (1995) claims that by working together and developing personal relationships we are able to develop an understanding of other members' risks and goals. This, he argues, allows us to develop trust and synergy based on understanding. Baden Hellard's message is entirely positive; however, we should recognise that the development of mutual understanding can be a major challenge for temporary project groups. Working together can also result in the development of distrust as we get to understand the motives and actions of others, well intended or otherwise.

Given that the majority of TPOs are set up for the duration of one project only it is highly likely that we find ourselves working with people that we do not know. Because we have no experience of these individuals we have no point of reference. This means that we have to rely on stereotypical opinions of others at the start of projects, which has been termed swift trust (discussed below). Over time, as we interact, we will have the opportunity to get to know our fellow group members, start to form opinions of others based on their actions, make judgments and hence start to establish a level of trust associated with the new relationship. If we find ourselves in TPOs that have the same members as a previous project then we do have an established point of reference from which to work – which can save time and effort. This does not mean that the previous experience was necessarily entirely positive, nor does it mean that the team members will automatically behave in the same way as they did previously as the context will certainly be different from the previous project. The point being made here is that we cannot take anything for granted, nor can we be complacent when it comes to human interaction. We still have to work at the relationships, albeit from a better-informed position than with an entirely new TPO.

The question of whether trust can exist within a team has been raised on several occasions. Jarvenpaa and Leidner (1999) investigated communication and trust in global virtual teams in which the members did not know one another before the start of the research. They found that trust can exist in global virtual teams that are built entirely on electronic networks. This challenges the views of Handy (1995), who questioned whether virtual teams could function without frequent face-to-face interaction, claiming trust needs the human touch.

Swift trust

Trust is not achieved instantly, rather it develops over time (Blau, 1964), and this can be challenging for temporary groupings of individuals and groups. There may be little time for trust to develop given the speed with which work has to be completed and so newly formed temporary project organisations have to rely on swift trust. Swift trust is a form of trust that is developed in temporary work groups (Meyerson *et al.* 1996). Members of the temporary project organisation usually have limited experience of working together, so they do not have a history of trust development, and in many cases they may have little prospect of working together again. The intensive time pressure associated with projects means that there is rarely time available for relationship development and so temporary groups tend to exhibit behaviours that presuppose trust (Meyerson *et al.* 1996). Because time pressures hinder the formation of opinions of other actors based on first-hand experience group members import expectations of trust rather than develop it, i.e. they form stereotypical impressions of others. So we expect architects to act in a certain way and engineers in another, and so on. Swift trust is formed when temporary teams are first assembled; thus according to Meyerson *et al.* (1996) swift trust is likely to be at its highest at the start of projects. As individuals start interacting they begin to gather experiences and form opinions that may reinforce or challenge the anticipated stereotypical behaviour and hence the level of trust. The longer the relationship continues the greater the opportunity for trust to develop.

Corruption and ethics

It is not possible to address trust without tackling the ever-present spectre of corruption, the potential for which exists in all business and project environments. All contributors to AEC projects, whatever their discipline, should act with integrity, i.e. everyone should be open, transparent and honest in all aspects of their business activities. Unfortunately, this is not always the case, with fraudulent and corrupt behaviour coming into the public domain from time to time. The recent exposure by the Office of Fair Trading of major construction organisations involved in cartel-type activity when bidding for public sector contracts in the UK helps to underline the amount of money involved. AEC projects usually involve extremely large sums of money and it is well known that money has the power to corrupt. Inherent characteristics of projects such as their size, fragmented nature, contractual links, uniqueness, complexity and the vast number of people involved also provide opportunities for the less scrupulous to make money through unethical, and often illegal, practices. The possibility of deceit and corruption exists in every projects and all contributors need to be vigilant.

Corruption takes two forms, bribery and fraud:

- *Bribery*. Bribery occurs when a representative of an organisation offers a bribe, either cash or a non-cash advantage such as a gift, to another party. This is sometimes done with the knowledge of the organisation's managers, sometimes without. Bribes are often paid through intermediaries or are disguised as a sub-contract arrangement in an attempt to conceal the transaction. The aim of a bribe is to gain advantage, for example securing a contract or approval for development that otherwise would not be forthcoming.
- *Fraud*. Fraud occurs when a representative of an organisation deliberately sets out to deceive another party with the aim of gaining some form of (financial) advantage. Well-known examples include contractors colluding to increase the contract price (called cover pricing) and contractors/sub-contractors surreptitiously substituting building products with cheaper alternatives. Other examples would include adding additional hours to time sheets to increase income and invoicing for materials not used.

Corrupt practices come in many guises and can occur at any stage within a project, some examples being bribing officials such as planning officers and members of the planning committee in an attempt to gain planning permission; collusion of contractors, manufacturers or suppliers to maintain high prices; paying bribes to win contracts; invoicing for work that was not done and for materials that were not used; bribing inspectors of the work to ignore poor work or unsafe working practices; pilfering of materials from site; and surreptitious product substitutions by a main contractor or a sub-contractor. Perceptions of what is, or is not, corrupt practice tend to vary between different countries, disciplines and individuals. This can pose a challenge for organisations working on international projects in countries where they are not familiar with what is, and what is not, deemed to be corrupt practice.

Although anecdotal reports of corruption in construction are widespread there is much less hard evidence to confirm or refute these perceptions. A report commissioned by the Chartered Institute of Building, *Corruption in the UK Construction Industry* (CIOB, 2006), aimed to identify the extent of corruption by analysing over 1,400 responses to a web-based questionnaire accessed through the CIOB's website. The results provide an insight into the problem, with the respondents reporting that corruption did exist, but the extent of it was not clear. Interpretation of what constituted 'corrupt practices' in relation to how the industry operated varied between the respondents, although the overall consensus was that more measures needed to be implemented to help prevent corruption.

Relational forms of contracting such as project partnering and integrated supply chains always carry the spectre of price fixing and so measures have to be put in place to prevent this from happening. Transparent and open accounting is one approach. Competitive tendering is one way of introducing an element of competition; however, it is not immune from misuse and corrupt practices.

Ethics

Closely related to personal and organisational values (discussed below) is the ethical approach taken by the organisations and individuals contributing to the project. Professional bodies and trade institutions promote ethical business practices in an attempt to eliminate corruption and give confidence to the public. Professionals are required to act with integrity, and guidance is given in their relevant professional code of conduct. For example, architects are covered by the Architects Registration Board (ARB) Code of Conduct, which sets out a number of standards that have to be complied with. If architects have joined a professional institution, such as the Royal Institute of British Architects (RIBA), they will be covered by the Code of Professional Conduct, which comprises three principles (integrity, competence and relationships) and a set of values that must be complied with. Although the wording of these codes varies between professions, the overall intention is the same, i.e. to give reassurance to the public and to fellow professionals. This is about trusting a professional to act in a transparent, honest and consistent manner and to look after the interests of their clients and society. If individuals step outside this set of values, i.e. act in a manner deemed to be unprofessional or unethical, then they are likely to be removed from the register. Reputable tradespeople are also governed by the rules and regulations relating to their specific occupation, also running the risk of expulsion if found to be operating outside the collectively held values of the trade body.

Corporate social responsibility

Corporate social responsibility (CSR) is a term used to describe how organisations manage their business affairs in such a way as to add value to society. In simple terms this means caring for the organisation's employees by treating them fairly and respectfully, while at the same time caring for the environment in which the organisation operates. Contributing in a positive manner to the environmental impact of construction could form an important part of an organisation's CSR; so too could the adoption of schemes such as the Considerate Contractor Scheme that operates in the UK. The challenge for organisations is to develop clear and unambiguous CSR policies that form part of their corporate values and culture, and which all of their members follow; otherwise it becomes little more than a corporate marketing activity and somewhat meaningless. Professional service firms are required to act in an ethical manner as codified in their profession's code of professional conduct, although some architectural practices have built on this in a positive manner, for example developing their CSR in relation to sustainability (Othman, 2009).

Values

An essential feature of interaction between project participants is the sharing of values, experiences and knowledge. Values are our core beliefs, morals and ideals, which are reflected in our attitude and behaviour and shaped through our social interactions. Our values, beliefs and attitudes are deep-rooted, part of our personality, helping to define who we are, how we interpret the world and how we interact with others, guiding our actions and decisions (Schwartz, 1994). However, our values are not absolute but exist in relation to the values held by others, and hence are subject to change over time.

Testing and confirmation of values is a positive characteristic of working across interdisciplinary cultures. Values are most commonly revealed through our conversation with others, i.e. they are expressed orally as we choose to express our opinions. Values are also subtly revealed by the ways in which we act and interact with others. In a TPO it can be beneficial when stakeholders discuss and establish a set of 'shared' values for a specific project, stage of a project, or series of projects (Austin *et al.*, 2001; Emmitt and Christoffersen, 2009). If client values are not fully understood it is likely to result in low fulfilment of client expectations and/or multiple design changes during the project, which can cause frustration and conflict within the TPO (Thyssen *et al.*, 2010).

Identification of values

Personal values are made up of behavioural (morals and ethics), cultural (meaning and traditions), political and religious (belief) values. Allport *et al.* (1960) classified values into six (now well-known) types in order to measure an individual's value orientation. Although individuals exhibit a range of values, each individual has a dominant orientation towards one of the following values:

- aesthetic
- economic
- political
- religious
- social
- theoretical.

Disciplines hold particular sets of values, which are related to their education and professional affiliation. Individuals and their organisations bring differing values, knowledge and interests to the project. Differences of opinion will emerge as interests and values are developed and challenged through interdisciplinary working. Usually the testing and confirmation of values is a positive characteristic of working across disciplinary cultures; however, sometimes the differences can result in disagreement and negative conflict. In simple terms there are three levels of values in TPOs:

- *Individual (personal) values.* These are influenced by our discipline.
- *Organisational values.* These are expressed in corporate mission statements and the way in which an organisation conducts its business.
- *Project (shared) values.* Collective project values are often implicit, emerging as the project evolves.

Although some of these values will become evident during the course of a project, it could be worthwhile exploring individual and organisational values at the start of projects, as these will influence how individuals interact with others. Selecting participants based on their values has proven to be effective for some AEC projects (Emmitt and Christoffersen, 2009). Individuals' values may differ from the values of the organisation they represent, and so a change in personnel during the project may have a major influence on the values applied by that organisation's representative(s).

At the level of the individual project it may be very difficult to improve working methods even when all participants and organisations agree to some common values. Maister (1993) has argued that many firms do not share values within the organisation and also fail to adequately discuss values with clients early in the appointment process. The implication is that the sharing of values is a challenge for organisations and temporary project groupings. The challenge is not exclusively with the implementation of tools to streamline the process, it concerns the interaction of organisations, or more specifically the efficacy of relationships between the actors participating in the temporary project coalition. This social interaction needs to be managed and someone should take responsibility for leading the process, as highlighted in the case study presented in Chapter 10.

Values-based process models seek to explore and agree values to aid development of the TPO and enable a smoother process. Discussion and sharing of values is largely achieved through face-to-face discussions within facilitated workshops. The workshops allow actors to discuss, explore and (hopefully) agree to commonly held values, often expressed in a written document as a set of value parameters and prioritized in order of importance to the project team. Working with shared values is a fundamental principle behind philosophies such as partnering and other forms of relational contracting. Agreed values are then prioritized in a list of value parameters, which forms the basis of a partnering agreement.

Many contractors rely very heavily on sub-contracted and sub-sub-contracted labour. So do architects and engineers, outsourcing non-core activities to other professionals in an attempt to stay competitive. It would be misleading to assume that the suppliers of sub-contracted labour share the same values as the main contractor. Similarly, it would be unreasonable to assume that the suppliers of sub-contracted work share the same values as the architects or engineers; they do not. Discussing and sharing values with the aim of establishing common project values early in the project is crucial to the successful development and delivery of projects. Recognising that differences of opinion may emerge as interests and values are developed and challenged, and hence values and priorities may change as the project proceeds, is also crucial. To recognise and respond to the values of others, and to align and reinforce the values of the project team, there is a need for effective communication skills.

Attitude

If we assume that it is a sensible policy to try to build and then maintain a high level of trust and integrity then we must be prepared to put appropriate tools in place and resource activities that encourage mutual understanding and respect. Assembling a TPO with organisations that share a similar vision and commitment to continual improvement is an important step in helping to deliver successful projects. This also means that the line managers within the organisations need to consider the attitudes of the individuals assigned to the project. Attitude is a settled opinion or way of thinking and this can be positive or negative in relation to a project. Individuals will tend to have either a positive 'can do' attitude or a negative 'cannot do' attitude. Most people, given a choice, would rather work with individuals who exhibit a positive attitude.

Value

Discussion of value can be traced back to Aristotle's *Nicomachean Ethics* (Korsgaard, 1986), and, although definitions vary, Perry (1914) has argued that value consists of the fulfilment of interests. Value is usually expressed in monetary terms and defined by economic measures, although value can also be seen more widely in terms of something's usefulness to an organisation. According to Menger (1950) value is formed subjectively by individuals, and hence value is a subjective measure that differs between individuals. Although value is a subjective judgment of something's worth or scarcity, it is common for value to be expressed in objective measures such as cost or price (e.g. value for money). Other factors relating to utility, aesthetics, cultural significance and market are also relevant measures of value, some of which are easier to quantify objectively than others. In a project setting value is what an individual or organisation places on a process (the project) and the outcome of that process (the product). Within this context the agreement of an objective best value for a group will differ from the individual's perception of value (Christoffersen, 2003).

As a subjective measure value changes depending upon one's circumstances and expectations at a particular point in time, i.e. it is dependent on context or the position of the person making the judgment. This is an important observation to make in relation to projects because individuals' context changes as the project evolves, thus there is likely to be a difference in perceived value between the start and the end of a project.

In the AEC sector the prime goal of a project is to deliver maximum value for the client while making a reasonable profit on the resources invested. From a client's perspective value is the relationship between price and quality, and tools are needed to measure value so that changes in value can be expressed. Thus it is common for value to be defined simply as function divided by its cost, which takes little account of philosophical factors.

Added value

Value for money is being able to do more with less. Construction clients are increasingly expecting more (better quality buildings) for less (capital investment). This expectation puts pressure on the members of the TPO to add value to the services they provide. Added value is a technique used for measuring organisational productivity and relates to the contribution (value) a process makes to the development of products or services. By reviewing the activities required to fulfil a business activity it is possible to identify activities that add value (value adding) to the business and those which do not (non-value adding). The non-value adding activities are wasteful and need to be eliminated. Value adding activities could, for example, relate to a new production process that saves a manufacturer money by reducing the amount of resources (time, labour, materials) consumed during production. Non-value adding activities could relate to the time spent waiting for other project participants to finish their work package, which could be eliminated by better coordination and planning of work. Value engineering techniques are used to add value for the client by seeking to reduce the cost of the building without compromising its quality. Similarly, lean thinking aims to reduce waste in the process and hence improve efficiency, thus adding value for the client.

Value to the end customer is an important aspect of the lean manufacturing philosophy (Womack *et al.*, 1991; Womack & Jones, 1996). Lean thinking and techniques borrowed and adapted from lean manufacturing can provide a useful array of tools through which the value of the design and the value delivered by the production processes can be enhanced. The lean approach helps to reduce waste in the process and can often be used to reduce waste in materials also. Although developed specifically for manufacturing and mass-produced products, the philosophy is relatively robust and can, with some interpretation, be applied to a project environment.

Managing value

Establishing value frameworks is an important principle behind integrated collaborative design (Austin *et al.* 2001). Similarly, the establishment of common objectives and common values are important objectives in the drive for greater cooperation and reduced conflict in construction projects (e.g. Kelly and Male, 1993).

If the value provided to a client is to be managed, efforts must be made to explore value as perceived by the client. Once this has been done it is then possible to express the client's value requirements as a number of value parameters that inform and shape the development of the project. Similarly the risks and uncertainty associated with the project must be identified and the consequences managed, otherwise value will be compromised.

Value management and value engineering techniques aim to articulate value for the project as perceived by the key project participants, in which value is related mainly to the cost of supply (economic value) and utility of the built artifact (use value). Aesthetics, cultural significance and market also feature in these discussions. Value management originates from value engineering, which was developed in the American manufacturing industry in 1947, and which spread to the AEC sector in the late 1960s (Kelly *et al.*, 2004). Value engineering strives for an optimum solution (required function at least cost). Value management acknowledges that an optimum solution may not exist, but the objective is to explore the challenge from a group perspective. Value management is often applied from the outset of the project through workshops in the early (conceptual) phase to determine the way ahead, whereas value engineering is usually conducted as an audit of the design.

The Japanese approach to value engineering focuses on continuous improvement throughout the process (Kelly *et al.*, 2004). This is perhaps best illustrated by the lean product development used by Toyota, which emphasises the importance of assessing many alternatives, a method known as set-based concurrent engineering, and the focus on elimination of waste in the entire development and manufacturing process (Morgan and Liker, 2006).

Managing risk and value

Underlying all projects is the amount of risk an individual and their organisation is prepared to take. This is primarily related to the amount of uncertainty and risk tolerance of individuals and their immediate managers. It is also related to the degree of trust that participants have in one another to perform their tasks. This is coloured by organisational culture and rituals as well as by the interaction with a diverse range of project stakeholders, some of whom will be more risk averse than others. Risks can be managed using a variety of risk management techniques and uncertainty can often be dealt with by clear communication and identification of roles and responsibilities.

Risk management should be linked to value management (Dallas, 2006). Risk management techniques aim to identify risks and uncertainty and mitigate their adverse impact on the project. Value management and risk management are complementary activities that inform the design team, and should be incorporated into the project framework. Apart from helping to maximise value and minimise risks the techniques help participants to develop a deep understanding of the project and develop a sense of ownership. By working together the opportunity to develop team culture and share knowledge through interpersonal interaction is also enhanced. Although it may be difficult to quantify all of the benefits of integrating value and risk management within the project framework it would appear that such efforts are instrumental in determining project success.

Both value and risk management techniques rely on the interaction of key project participants to discuss the value and risk associated with the project at key stages, which is usually done in facilitated workshops. The number of workshops will be determined by the size and complexity of the project and the attitudes of the key participants. As a minimum workshops should be held at the start of the project (inception and client briefing), at the interface between design and realisation, at commissioning and during use.

Practical challenges

Many of the issues highlighted in this chapter are often hidden from managers and fellow project participants. Although organisations are keen to communicate their CSR and their corporate values through their home pages and associated marketing material, the positive aspects of how they work, the less positive aspects of business rarely come into the public domain. It is not an easy task knowing whom to trust, or for that matter what to trust them with, until we have had some experience of working with previously unfamiliar organisations and individuals. Individuals are vulnerable when entering new relationships, but unless we expose ourselves to some risk there will be little chance of improving working methods.

Conflict is most likely to manifest when individuals are uncomfortable with their project identities, i.e. they do not share project values. Therefore, it is important that participants understand their own and others' roles and identities. As already mentioned, this can be challenging as many contributors may have hidden agendas for participating and their sporadic interaction with others may make it very difficult to establish common ground. Facilitated workshops can be an effective means of bringing people together to discuss their roles in a project before the project starts, or at key stages in the project (e.g. the transition between design and construction).

Jealousies and personal rivalries exist within organisations. It would be unreasonable to assume that they do not exist within temporary project organisations – they do and they need to be managed with sensitivity.

One of the most challenging aspects of a manager's job is to try to get an understanding of the trust levels and values of those participating. This is rarely a skills set taught in university education programmes and it can be a daunting area for managers whose primary skills are technical, rather than social.

End of chapter exercises

- Identify a person whom you trust and list up to five reasons why you trust them. Now try and think of an event, situation or reason that might make you question, and hence reconsider, your attitude towards that individual.
- Identify a person whom you do not trust and try to list up to five reasons why you do not trust them. Now try and think of an event, situation or reason that might make you question, and hence reconsider, your attitude towards that individual.
- You have been asked to manage a new project by your line manager. The main project participants have already been selected by the client and have not worked together before. What would you do to develop trust within this project grouping?
- Compare two codes of professional conduct (e.g. Architects, Architectural Technologists). What characteristics do they have in common?

Further reading

Value and Risk Management: a guide to best practice by M. F. Dallas (2006) provides a comprehensive guide to the management of both value and risk. Further insights into trust can be found in *Trust: the social virtues and the creation of prosperity* (Fukuyama 1995).

4 Discussions

Discussions with our work colleagues form an essential part of our working day; they help us to establish information and help to provide the context for making informed decisions. Discussions are also used to help build and reinforce relationships within the workplace, test ideas, negotiate and reinforce mutual trust and respect. Many of the informal discussions are relatively short-lived, lasting a few seconds or minutes, and are soon forgotten; they are simply a means to an end. By comparison, more formal discussions in meetings tend to last for longer and can have far-reaching consequences for the performance of the organisation and project alike.

Within the office informal discussions, either on a one-to-one basis or in small work groups, allow individuals to share knowledge and test ideas in a supportive, psychologically comfortable and 'safe' environment. The language used is familiar to those participating and the context is shared, allowing communicators to discuss issues in a relaxed and informal manner. A good example of this is the architectural design studio, which is planned so that the layout of the office encourages *ad hoc* interaction, facilitates informal group work and stimulates informed decision making. This encourages communication with the aim of stimulating creativity and effective group work. Interactive media provide the opportunity to work concurrently with other individuals and work groups in co-located offices.

In a project environment face-to-face social interaction will usually occur in meetings, of various types and formality, and workshops, which by their very nature tend to be relatively informal. Meetings are usually held in one physical location, such as the architect's office, the offices of the town planning department or within the contractor's site accommodation, often consuming considerable amounts of time in travelling to and from the location of the meeting. It is becoming common for telephone and videoconferencing and/or online meetings to be used for their convenience to the participants. In the majority of projects there will usually be a mix of face-to-face interaction and online meetings. Workshops rely on close interaction between participants and because of this they are conducted in a space that allows those taking part to work closely together to develop their relationships.

Meetings and workshops are essential mechanisms for ensuring effective communication, information sharing and discussions, and facilitating decision making. This allows project participants to discuss issues and agree on the most appropriate action. Meetings are held because people who have complementary functions within the project need to communicate to accomplish a common task. Workshops fulfil a similar function, although they are often used to develop social interaction between team members while addressing a particular task, i.e. they serve two complementary functions, problem solving and developing interpersonal relationships to encourage integrated working. Workshops are a common feature of value management exercises.

Fundamentals

A factor affecting social interaction is the power of each side to affect the other. Influencing skills enable the development of relationships and are a key skill of good managers. People with a high persuasive ability can use their skill to handle conflict constructively and hence to promote openness and constructive debate. To perform effectively construction project managers need the ability to negotiate and persuade others to take action. One way to view negotiation is as a struggle, in which concealment and competitive tactics feature strongly. An alternative view of negotiation is as collaboration, a process in which parties make sacrifices rather than demand concessions in the pursuit of some (mutually beneficial) overriding project goal. When parties use collaborative strategies during negotiations it is more likely that a mutually agreeable solution will be produced. In situations in which negotiators or project managers feel that they have to satisfy tough demands (e.g. negotiations with stakeholders within a project) they will adopt a more competitive approach. Both the collaborative and competitive styles of conflict involve positive and negative socio-emotional discussion. The collaborative method would have a tendency to maintain and repair relationships during discussion. The competitive style would be prepared to threaten the relationship and have a greater tendency towards extremes of the negative socio-emotional traits. Both approaches would also use task-based logic to explain the rationale; however, getting mutual understanding would be far more important in the collaborative approach than in the competitive approach.

Rational persuasion is often used during negotiations. The most common form of rational persuasion consists of logical arguments and factual evidence to help develop understanding and explain the situation. Rational persuasion is most appropriate when the opponent shares the same task objectives, but does not recognise the proposal in its current format as the best way to attain the objectives. In a situation in which the opponents have incompatible objectives or end goals, this type of influencing tactic is unlikely to be successful for obtaining commitment or partial agreement.

Disagreement and agreement

Disagreement is usually seen as a negative term, yet it is found in most group interaction. A certain amount of challenge, evaluation and disagreement is necessary to appraise alternatives and reduce the risks. Emotional expression helps others to recognise an individual's preferred beliefs, behaviours and actions. It exposes the structure and routine in which the individual would like to work. In groups such expressions are needed to establish what is acceptable and what is not and in which areas conflict is likely to occur. The decision-making structure of a group is dependent on individuals expressing their agreement and disagreement and working through any variance. As individuals work through their differences they develop a much deeper understanding of the beliefs and values of others. However, people may choose to avoid disagreements to enable them to pursue relationship goals, believing that disagreeing would weaken the relationship. Clearly it is important to expose differences of opinion and explore issues in depth, but it is equally important to identify and respond to an individual's values and beliefs.

During difficult tasks and stressful situations, members of the group are more inclined to pursue relationship goals, supporting each other rather than dealing with the problem and enquiring about the risks involved. Pressure to agree may be so strong that group members may continue to agree blandly while unwittingly consenting to their own destruction. Such attributes are associated with groupthink. Groupthink occurs when members of a group do not agree with statements that are made but do not make their view known to others, which results in the group members believing that agreement is reached. Ways of avoiding groupthink include asking questions, noting an absence of disagreement (which serves as a warning to group members to reassess alternatives) and being aware that the risk of illusory agreement heightens as external stress increases. Seemingly unanimous agreement by the group may disguise a silent minority.

Defensiveness and barriers

Defensiveness can be defined as the behaviour of an individual when she or he perceives threat in the group. A person who perceives threat may communicate in a guarded or attacking way. Defensiveness of this nature will manifest behaviour patterns that are either consciously or subconsciously recognisable to other parties. The inner feelings of defensiveness create outwardly defensive postures. If such actions take place without question an increasingly circular destructive response may occur. Defensive signals are said to distort the message. When a receiver attempts to understand a communicated message they also extend their efforts towards understanding the motives behind defensiveness, attempting to understand why someone is behaving in this manner. Excessive defensive and aggressive behaviour distracts those engaged in the interaction away from rational discussion. Defensive arousal prevents the listener from concentrating upon the message. The defensive behaviour increasingly distorts as the circular defensive behaviour continues. The conversation moves from the subject matter to defensive action and reciprocal attacks. Defensive and supportive communication can be accommodated without changing the content of the statement. For example, enquiring rather than demanding information reduces defensiveness and aggressive responses. Good communicators develop flexible interaction techniques that enable a greater appreciation of the other's perspective.

Help seeking and question asking

Help-seeking behaviours are fundamentally interpersonal, i.e. one person seeks assistance from another. Individuals are more likely to seek help from equal status peers and others who have helped them earlier; cooperative patterns are reciprocal. Help-seeking behaviour implies incompetence and dependence, and many professionals are reluctant to ask questions for fear of being perceived in such a manner. It is likely that high status professionals will avoid situations in which they need further information in order to avoid asking questions and to defend their status. Serious and costly errors have been made in multidisciplinary projects that could have been prevented by seeking expert help that was available at the time.

Meetings and workshops

To be an effective management tool meetings and workshops must have a clear purpose, i.e. they should be productive. The objective of the meeting or workshop must be clearly defined and the expected contribution of those invited to participate must be clearly expressed. If the objective can be achieved only by bringing certain people together at a certain time, either physically or virtually, then it is necessary to hold a meeting or conduct a workshop. In this case only the individuals considered necessary to achieve the objective of the meeting or workshop should be invited. If the objective can be achieved by an alternative (cheaper and/or quicker) means, such as writing a letter, sending an email or making some telephone calls, then the alternative should be used.

The productivity of meetings and workshops is an important consideration for a number of reasons. Both forums consume a considerable amount of the participants' time in preparing, travelling and attending. Understandably, they should be used sparingly and be integral to the project's strategic process plan. Meetings and workshops should be managed and facilitated in such a way as to encourage effective decision making. A meeting would be deemed productive if the aims and objectives have been met within a set timescale. The value added to the project is the ability to encourage interaction and candid discussions with a view to making informed decisions. Although a variety of metrics could be used to measure the efficacy of meetings and workshops the main concern for managers should be with the process (effectiveness of group processes and communication), the product (quality of the outcome/solution) and perception (how the participants perceived the process and the product).

Internal and external events

From the organisation's perspective meetings and workshops can be classified as one of two types, either 'internal' or 'external' to the organisation (or department). This can have a significant bearing on how individuals communicate and how trusting they are of others.

- Internal (closed) meetings and workshops are limited to participation by the organisation's members only. In this familiar environment it is possible for the participants to be relatively informal, open and trusting of the other members. Discussions tend to be relatively open with shared objectives. An example of this type of interaction could be a number of designers discussing the design development of a particular project within their office.
- External (open) meetings and workshops include the presence of participants from other, possibly competing, organisations or departments. In this environment individuals may be expected to act in a more formal manner and be less candid and less trusting of the other members' motives (regardless of procurement route). Discussions tend to be relatively guarded and objectives might not be shared. An example of this type of interaction could be key members of a project team discussing the progress of a particular project in the contractor's site accommodation office. In this example, progress of the work will tend to influence the communication behaviour of those present in the meeting. When things are progressing as planned participants tend to be relaxed and relatively open in their communications. However, when the work is behind schedule and problems are known to exist participants have a tendency to be less relaxed and relatively closed in their communications, at a time when a candid exchange would most probably be more productive. With regard to workshops, one of the early tasks of the facilitator is to promote social interaction through various 'ice-breaking' exercises to try and encourage participants to relax and participate in a less guarded manner.

When viewed in the context of a project the majority of meetings and workshops will involve individuals from a variety of organisations and will be external to the individual and his or her organisation.

Time and place

Consideration should also be given to the location of the meeting or workshop because individuals and groups are known to behave differently in different physical settings. For example, trying to hold an impromptu meeting on a wet and windswept construction site will foster different behaviour (and possibly outcomes) from holding the meeting with the same members in a warm and comfortable site office. In such a situation it may be necessary to analyse a particular part of the building and discuss it *in situ* in an attempt to resolve problems, and this often has to be done regardless of the prevailing weather conditions (assuming it is safe to do so). Then the participants can retire to the site offices to discuss the options and agree how to proceed. We also tend to behave differently when hosting a meeting or workshop in the comfort of our own offices from when attending an event hosted by someone else in unfamiliar surroundings. Care is required in selecting the venue for a meeting when discussing contentious issues; a neutral location may be beneficial to discussions.

To a certain extent the use of mobile technologies has helped to reduce the number of visits to site for relatively minor issues, but sometimes a physical presence is necessary to discuss technical and/or sensitive issues face-to-face.

Meetings

There is no escape from the ubiquitous meeting. Mintzberg (1973) found that managers spent nearly seventy per cent of their working day in meetings, the majority of which were scheduled and the rest impromptu. This finding tends to be supported in subsequent studies (Hartley, 1997). However, despite the amount of time and energy consumed by meetings they are relatively under-researched within the management literature (Volkema and Niederman, 1995) and also within the construction literature (Emmitt and Gorse, 2007). Although a small number of books have been written mainly by practitioners with the aim of providing guidance, Hartley (1997) concluded that we do not really know what goes on within this forum, despite their familiarity.

Research into construction progress meetings (Gorse, 2002) found that the meetings were an important tool for helping the AEC project to progress in a positive manner. Discussions within the meetings were focused on a specific task or problem solving and not surprisingly most of the communication was task-based. Participants also used socio-emotional communication to disperse tension and help build relationships. Gorse found that there was a correlation between well-managed and chaired meetings and project success, with the most successful project managers being better at chairing meetings and steering discussions than their less successful peers.

Function

In addition to facilitating the exchange of information, ideas and opinions and providing a forum for decision making, meetings are also used to:

- *Appraise*. Meetings are used to appraise progress and the performance of projects, organisations and individuals.
- *Bond*. Meetings fulfil a fundamental human need to communicate and bond, and hence help foster team relationships. They create a sense of belonging and reflect the collective and cultural values of the temporary project organisation. Meetings can also be used as a tool to help motivate the project team, although this function may be better served through facilitated workshops as discussed below.
- *Control*. Meetings allow managers to stay apprised of progress and in command of the tasks to be completed. They also allow those attending to follow up information requests, allocate scarce resources, agree action and set deadlines. All decisions should be recorded in the meeting minutes.
- *Coordinate*. Face-to-face discussion may help with the coordination of work packages and the clarification of roles and responsibilities. The aim is to ensure that adequate resources are allocated to allow operations to take place effectively and safely.
- *Develop trust*. Addressing tasks and resolving problems in a meeting forum can help to develop trust between individuals as others are found to act with benevolence. Conversely, the failure of individuals to engage fully in problem solving will usually lead to (or confirm) a lack of trust. Either way, it is useful to know where the boundaries of trust lie.
- *Explore possibilities and preferences*, for example through structured client briefing exercises.
- *Resolve and clarify*. A timely meeting can help to resolve problems, differences of opinion, minor conflicts and disputes. It can also help to clarify certain aspects that might subsequently have led to unnecessary errors and rework. This may be something as simple as misunderstanding how words are used or seeking clarification about apparent differences between the drawings and the written specification.

Meetings should not be considered as isolated events where decisions are made; instead they need to be seen in a wider context. This includes the incremental cycle of social interaction that is used to share and process information, make and confirm decisions and develop and maintain relationships. Many different types of meeting are convened during the life of an AEC project to serve a variety of complementary functions. These range from the informal to the formal, and the impromptu to the strategically planned. The main reason for holding a meeting is to bring people together in one place (physical or virtual) to discuss issues, and they can be used to:

- start projects
- develop and maintain effective teams and groups
- explore values and agree value parameters (e.g. briefing)
- discuss and review progress (e.g. of the design, the project)
- discuss and resolve disagreements
- exchange information and knowledge
- discuss and resolve problems
- close projects
- hand over projects (or stages of projects)
- analyse projects (e.g. to gauge performance of the participants).

Making meetings work

Whatever framework is chosen for a project it will involve a number of meetings, and these need to be planned to be effective and efficient. Meetings must serve a function, i.e. they must add value to the process, otherwise they contribute to process waste. There is little point in holding a meeting for the sake of it, and so all scheduled meetings should be reviewed to check that they are still required. Equally there is little point in holding a meeting without clear aims and objectives. The vast majority of us would readily complain that we have to attend too many meetings and hence do not have sufficient time to do our work. This reaction is grounded in the reality that too many meetings are arranged without pausing to ask whether they are necessary, and too many are poorly organised and chaired, hence they are wasteful. When meetings are strategically planned (e.g. as part of the project plan), well organised and appropriately directed, they have been proven to be a major benefit to the progress of projects and the development of organisations.

Chairing the meeting

Research by Emmitt and Gorse (2007) found a positive relationship between project success and the ability of project managers to chair meetings effectively. A good chairperson avoids competing with others, encourages everyone to contribute and is able to control aggressive and defensive behaviour. Furthermore, he or she should be able to summarise decisions concisely and clearly, summarising the decisions made, stating the agreements made and timescales for completion and identifying those responsible for the action. A good chair will also set an appropriate pace for the meeting, allowing individuals the opportunity to contribute and challenge, without dominating the discussions.

Participation and interaction

There are no guidelines on the number of individuals who should participate in a meeting, although as a general rule the larger the number of attendees the less the potential for all to participate fully in the discussions. It is common to find that AEC contracts require representation by the main contributors at regular progress meetings, which can mean that the meeting has between ten and twenty participants. Obviously, the greater the number of individuals involved the more difficult it is to ensure that everyone participates, thus it may be necessary to phase participation within the meeting to ensure that all views are represented.

Interaction during meetings is not the same as in casual conversation. Meetings normally have an organisational purpose and restricted turn-taking procedures. Furthermore, interaction is often hierarchical; the more senior members of the TPO tend to take the position of chair and often assume greater participation rights, although turn-taking is rarely fixed. Meetings also develop norms, and not all participants will be comfortable discussing issues if the norm is overly oppressive or skewed in favour of certain participants. The onus is on the chair to ensure a comfortable and fair environment for discussion and decision making.

Use of interactive media

Interactive media can be used in meetings to help reduce the amount of time spent describing issues, hence increasing the amount of time dedicated to discussing pertinent issues. Computer-mediated workspaces, which include large interactive screens and laptops, are being promoted as a means of improving communication and information coordination. Virtual prototyping provides the technology to discuss designs, constructability and the scheduling of construction work in large immersive laboratories. Access to these facilities is currently a barrier for the majority of projects as the facilities tend to be few in number and located at universities and research institutions.

Managing the meeting

There are some simple rules to follow that apply to face-to-face meetings and meetings held at a distance. These are described as a series of steps.

Step 1 – Define the aims and objectives of the meeting

Before convening a meeting it is necessary to address some fundamental questions:

- What is the purpose of the meeting?
- Is a meeting the best way of dealing with the issue(s)?

- Who needs to attend/participate and what is their expected contribution?
- How many individuals need to attend (participate) from each organisation?
- How long does it need to last?
- Do participants need to meet in one physical location or can the meeting be conducted at distance by telephone or videoconferencing?
- Is this a one-off meeting, or is it part of a series of meeting to discuss related issues?

In attempting to answer these questions it will be possible to clearly state the purpose of the meeting and identify, and subsequently invite, only those individuals best suited to dealing with the issue(s) to be addressed.

Step 2 – Issue invitations and the agenda

Once a suitable date and venue has been agreed it is necessary to issue invitations. It is good practice to send out an agenda for the meeting at the same time as the invitation to attend. The agenda should set out the items to be discussed and identify those responsible for providing brief reports. The agenda should also give clear time frames for the meeting and the chair should endeavour to stick to the agenda.

Step 3 – Open the meeting

The meeting should start on time. If one or two participants have failed to arrive on time, start without them. (Next time they are more likely to be punctual.) The chair should open the meeting by outlining the aims and objectives of the meeting and emphasising the need to stick to the time frame. Participants should be introduced and their ability to take decisions on behalf of their organisation confirmed. It is not uncommon for an individual to deputise for a colleague, but not have the responsibility to take decisions (which is usually a waste of everyone's' time).

The frustrating (and common) ritual of reading through the minutes of the previous meeting should be avoided. It is unnecessary and wastes precious time.

Step 4 – Discuss issues in turn

A good chairperson realises the importance of allowing all participants to contribute to the discussions. This may mean that some effort will be required to encourage reluctant communicators to participate and also to limit the contribution of the more eager participants.

Step 5 – Reach a decision and confirm it

The aim is to reach a decision or, failing that, to reach agreement that more information is required before an informed decision can be made, i.e. delay a decision. In either case it is necessary to confirm what was agreed and if necessary allocate responsibility for the action necessary to complete the task. When possible a time frame for completing the work should also be agreed and recorded in the minutes of the meeting. Combined, this makes it simple to track decisions, responsibilities and time frames for action.

Step 6 – Close the meeting

At the close of the meeting the chair should take a few moments to summarise the decisions made and reiterate responsibility and time frames for decisions. It is good practice to confirm the date of the next meeting (if appropriate) and highlight critical dates for submitting reports prior to the next meeting. Participants should then be thanked for their contributions.

Step 7 – Write and circulate the minutes

It is common for the chair of the meeting to write the minutes of the meeting. These should be circulated as soon as practicably possible after the meeting, usually within three working days, while the meeting is still fresh in the minds of the participants. Minutes should concisely and accurately confirm the main thrust of the discussion, decisions, actions and responsibilities for carrying out commitments by an agreed deadline. Participants should check the minutes to ensure that they represent an accurate record of the meeting. If there are any discrepancies the author of the minutes should be contacted immediately (do not wait until the next meeting, by which time it may be too late to reverse a decision).

Decisions outside the meeting

Many decisions are made outside the meeting forum, either before it starts, in discussions during refreshment breaks, or after the closure of the meeting. These tend to be face-to-face discussions between two or three individuals anxious to reach consensus over a particular issue in order to present a united view and thus help avoid uncertainty, disagreement and conflict within the meeting. Some managers will ensure that participants have adequate time to meet before the start of the meeting, for example by walking around the construction site to discuss progress. By doing this participants can discuss issues informally, which may help to save time in the meeting, but which also helps participants to gain a better understanding of one another. Similarly, refreshment breaks should be long enough to allow individuals to reach agreement over contentious issues, should they arise. Arranging lunch or light refreshments after the close of a meeting can also be a useful way of allowing participants to discuss the outcomes of the meeting informally, before they return to their offices. Although these informal events will consume additional time, experience would tend to suggest that they can significantly reduce the amount of time spent in the meeting trying to reach understanding, and hence they might not constitute wasted time.

Dysfunctional meetings

Not all meetings will achieve their stated aims and objectives and in some cases a meeting may have negative outcomes. Sometimes an unexpected event or information comes to light during the meeting, causing the focus of discussions to change. Occasionally there may be a clash of views or personalities, leading to disagreement and frustration. Alternatively, the meeting can add little value if not chaired in a professional manner. When meetings become dysfunctional valuable time can be wasted, and in situations in which the meeting has led to ill feeling or frustration trust and communication routes can be damaged. There are a number of root causes that need to be addressed:

- Participants ignore group knowledge. Individuals often fail to build on the ideas of others, waiting instead for an opportunity to project their own ideas and ignoring other discussions.
- Concentrating too much on one train of thought, at the expense of other items on the agenda.
- Too much time is wasted discussing irrelevant issues and hence too little time is devoted to the major challenges (which may make some participants feel uncomfortable).
- Failure to ensure that all those present contribute to the meeting.
- Failure to explore alternative views and options (leading to groupthink).
- Failure to devote time to socio-emotional interaction by being too task-focused, thus relationships are slow to develop.

Facilitated workshops

The majority of meetings are convened to deal with procedural issues and are primarily concerned with the achievement of tasks. Interaction is mainly task-based. Workshops provide a forum for creative interaction and have a central position in collaborative design approaches and the development of integrated temporary project organisations. Workshops differ from meetings in that they are concerned with establishing and developing interpersonal relationships, as either a primary or a secondary function of the workshop. Interaction is mainly socio-emotional. Development of relationships is often achieved by working collaboratively towards solving a (non-project-specific) task (e.g. a simulated role play exercise or an educational game) or by working collaboratively on a project-specific issue, for example in a value management (or value engineering) workshop.

Function

In addition to helping to establish group membership and social identity in a temporary organisational setting, facilitated workshops are also used to:

- build trust
- confront groupthink
- create knowledge
- develop working relationships
- establish project parameters
- explore different perspectives (and disagreements)
- resolve conflict.

Compatibility and values are difficult to establish from CVs and personal recommendations. It is not until people start to work together that the level of compatibility starts to become clear and values start to emerge through actions. Exploring the degree of compatibility through facilitated workshop exercises can be a highly effective way of accelerating individual under-standing of fellow contributors' values and preferences for communicating. Organisations may use 'awaydays' to allow their employees to understand one another better through non-work-orientated activities in a workshop setting.

Workshops are a common feature of value management and value engineering exercises. The workshops provide a forum for systematically evaluating design proposals and exploring alternative solutions that may offer improved value to the client and project stakeholders. By bringing outsiders into the project it is possible to review the project objectively. Workshops also form the basis for values-based process models, one example of which is described in Chapter 10.

Making workshops work

Workshops, like meetings, need to be structured to achieve a specific aim. They should, when appropriate, be included in the project plan, and their aims and objectives should be clearly stated. Unlike meetings, workshops tend to be used sparingly, often at the front end of projects to stimulate teamwork and develop trust, or at specific points in the project as part of a value management framework (see Kelly *et al.*, 2004).

Guidelines on the number of people who should attend facilitated workshops tend to vary. The important point to consider is the number of individuals that the facilitator (or facilitators) feel comfortable with and the context in which the workshop is set.

Facilitating workshops

Facilitators tend to have individual approaches to starting workshops, sometimes called 'icebreakers' or 'getting-to-know-you' exercises, which are primarily aimed at getting the participants to start interacting. Techniques vary, but this usually involves the participants being asked to introduce themselves and describe their role within the project or organisation, their qualifications and experience and some personal information, such as describing their favourite hobby. This is then followed by a simple and quick group task designed to get people working together, usually in small groups of three to four people, following which the small groups are encouraged to discuss the outcomes. Once participants have started to interact and relax then it is possible to move on to the main purpose of the workshop.

Facilitators have a significant role to play in creating an environment in which participants are happy to communicate openly and contribute to discussions candidly. To a certain extent this is determined by the personality of the facilitator and his or her ability to bring about a trusting atmosphere very quickly. This is discussed in greater detail in Chapter 10.

Use of interactive media and gaming

Playing educational games is a well-rehearsed technique for developing interpersonal relationships. Interactive media provide additional opportunities to engage workshop participants in a collective task. Some facilitators will use toy bricks, for example Lego, as a means of stimulating the participants to work together. Games based on solving problems and developing design solutions serve a similar function: to encourage socio-emotional development.

Combined meetings and workshops

In the creative phases of projects it may be desirable to hold combined meetings and workshops. The purpose is to review progress and identify areas that would benefit from collaborative input in the meeting part, and to break out into small groups to tackle some of these problems in the workshop part. Once these have been completed the groups then report back to the meeting, for example presenting a series of options for further discussion (and decision). This technique is sometimes used by architects and engineers keen to explore a number of options while the key project stakeholders are gathered together. Some contracting organisations also use this format to explore a number of options for addressing constructability and scheduling issues. Sometimes the workshop elements are facilitated, although it is more common for them to be managed by someone internal to the organisation, which is practical and cost-effective. A variety of creative problem-solving techniques may be used to suit particular circumstances (see Chapter 5).

Practical challenges

One of the biggest problems with both meetings and workshops is that of overuse and hence complacency. Meetings are perhaps the most obvious candidate for being overused, and hence the tendency is that participants give meetings less attention than they deserve, thus the meeting often becomes ineffective and little more than a ritual. Participants usually attend meetings because they are contractually bound to do so, or they feel a duty to attend, rather than because their presence is essential. In such situations it may come as little surprise that individuals may not contribute fully to the discussions (which is not always easy to spot).

Workshops can also become rather tiresome, especially when confronted with very similar approaches by the facilitator each time. Thus the tendency is for individuals to attend and simply 'go through the motions' or to make their excuses and not attend at all.

To overcome these potential challenges it is necessary to establish an appropriate framework for each project and review the need for meetings and workshops at regular intervals. This is best done in discussions with the participants to ensure a collective decision. Using an appropriate process plan for the project context, such as the RIBA Plan of Work, will help to establish the most appropriate opportunities for meetings and workshops. Bringing people together (both physically and virtually) is expensive, therefore the project manager must be able to justify the use of meetings and workshops as an integral and essential part of the project plan. Ensuring that the most appropriate people attend each meeting and workshop, making the event relevant to the project, and ensuring that decisions are made, communicated to the appropriate parties and acted upon will help to reinforce the value of both meetings and workshops.

End of chapter exercises

- You are planning a project with a duration of twelve months from inception to completion. How do you decide how many meetings to include? What factors do you need to take into account?
- You have been invited to a monthly project meeting. Last time you attended the meeting it was poorly chaired and you felt it was a waste of your time and you are unsure whether or not you should attend. What do you do to resolve your dilemma?
- Several months into a large project it is evident that there are some personality clashes within the temporary project organisation, which appear to be hindering communication and the ability to make decisions in a timely manner. Your immediate line manager thinks that a workshop would solve the problem. Discuss.

Further reading

Comprehensive guidance on value management techniques can be found in *Value Management of Construction Projects* by Kelly *et al.* (2004).

5   Decisions

Professionals are paid to make informed decisions on behalf of their clients. Indeed, it could be argued that the value added to the project is related to the quality of the decision-making processes. Understanding how individuals and groups make decisions and the pressures brought to bear on the decision-making process is an important factor in the successful management of interdisciplinary projects. Decision making in AEC projects involves a wide range of participants, making interdependent decisions at different times and at differing levels within the project. Communication skills and the ability to cooperate are essential in order to make decisions and hence achieve the various work packages. However, as noted earlier in the book, neither communication nor cooperation skills should be taken for granted; both require continual effort from the participants to ensure that the most appropriate decisions are made for a given context. Participants have varying levels of power and input during the project life cycle. Similarly, the values held by individuals, their trust in others and the ability to interact harmoniously (or not) will colour the decisions made. Furthermore, individuals respond differently to various decision-making stimuli and may wish to tackle problems in different ways to their fellow participants. This can have a bearing on the ability of multidisciplinary groups to make decisions and hence function effectively.

The manner in which the building performs in use, the options for modification and reuse, and the potential for recycling and minimising waste at disposal are determined to a greater or lesser extent in the conceptual design phases and coloured by decisions made during detailing and realisation. Underlying all issues concerned with design, manufacture and assembly is the ability to make decisions in the available time. Regardless of the building type, size and complexity of the design, each project will have some form of time constraint imposed on it. Usually, the client requires a completed building for a particular date, a date that will influence the amount of time allocated to different phases of the project. This imposes time constraints that have to be accommodated into overall programming of resources, thus limiting the amount of time available for discrete work packages. To help identify critical dates for making decisions it is necessary to develop a project plan and establish clear responsibilities and timescales for decisions. Some of these responsibilities will relate to individual work packages, others to multidisciplinary group work.

Decision making will also be influenced by the type of problem being addressed and the individuals involved. Because designers and engineers will face design problems that are ill-defined, poorly described or diffuse in nature, attempts must be made to define the problem clearly before it can be resolved. In some of the design literature, this process is described as 'questioning' (e.g. Potter, 1989). Definition of problems is made easier through the designer asking questions of himself or herself, and also of others. The aim of this questioning process is to be able to take full account of the information, explore possibilities and recognise the limitations, essentially a process of simplification. By comparison, the problem confronting the site worker may be clearly defined and visible (e.g. the lintel will not fit within the opening), although it is not always easy to find an economic and safe solution quickly.

Fundamentals

Much of the literature dedicated to decision making is centred on the actions of individuals, with less emphasis placed on the collective efforts of interdisciplinary groups. Clearly, it is easier to observe the behaviour of individuals, especially in controlled experiments, than the group activity of, for example, a design office or a project progress meeting. Therefore, this bias in the data available is to be expected. In practice, individuals are constrained and influenced by the behaviour of the group to which they belong and by interaction with other groups that are party to the project. There are also cultural constraints, which can vary within and between international boundaries. When people make decisions they tend to follow rules and/or procedures that they see as appropriate to the situation (March, 1994). This is particularly so of professionals who are expected to act in a manner appropriate to their particular professional background. For example, not only do architects have to satisfy their client, but also they have to satisfy different building users and, as in many other professions, will feel a need for peer approval.

Professionals are expected to act in a logical manner when solving problems, assessing all of the options against a background of legislation before making a decision. Nevertheless research suggests that this may not be the case. Although much of the literature on decision making makes assumptions based on rationality, the validity of this assumption is thrown into question by studies carried out by behavioural scientists. Observational studies of decision-making behaviour suggest that individuals are not aware of all of the options, do not consider all of the consequences and do not invoke all of their preferences at the same time. Rather, they consider only a few options and look at them sequentially, often ignoring some of the available information (March, 1994). Decision makers are also constrained by incomplete information and their own cognitive limitations, which are affected by factors such as age, illness, attention span, tiredness, stress and burnout. Thus, although decision makers may set out to make rational decisions, in reality they make decisions based on limited rationality: they search for a solution that is 'good enough', not the 'best possible' solution.

Human decision making

The factors that place constraints on human decision making are (March, 1994; Emmitt and Yeomans, 2008):

- *Attention span.* It is impossible to deal with everything at once. There are too many messages, and too many things to think about. Thus, we tend to limit our attention to one task at a time, ignoring messages that are irrelevant to that particular task, engaging our selective exposure. Our attention span is also limited by time, pressure brought about by work, and our private lives.
- *Memory.* Both individuals and organisations have limited memories. Our memories are not always accurate; we tend to remember acts as we like to see them (a characteristic known as constructed memory), rather than as they actually happened. Individual memories are usually classified as being short term and long term. Organisations and individuals are limited by their ability to retrieve information that has been stored. Records are often not kept, may be inaccurate or are lost so that lessons learned from previous experience are not reliably retrieved. Moreover, knowledge stored in one part of an organisation cannot readily be used by another part of that organisation.
- *Comprehension.* Despite having all of the facts to hand, the relevance of information may not be fully understood. There can be a failure to connect different parts of information. Furthermore, individuals have different levels of comprehension, making it difficult to foresee how each will respond to the information that they have. For example, the architect, manufacturer and contractor may understand the same piece of information differently, simply because of their different backgrounds.
- *Communication problems.* Specialisation, fragmentation and differentiation of labour encourage barriers and present difficulties in the transmission of information and knowledge. Different groups develop their own frameworks and language for handling problems, and communication between these cultures can become difficult.

One way of reducing the effect of these difficulties, while reducing the time required to make decisions, is to use familiar solutions. Designers draw on experience (their own and that of others) to come up with a particular design solution for a specific site and a particular client. Similarly, contractors and tradespeople also draw on their previous experience of using certain tools and products to achieve a task. Relying on personal experience requires a considerable amount of knowledge of various solutions to problems, knowledge that is only acquired with experience. The tendency is for young practitioners (with limited knowledge) to rely on techniques and solutions suggested by others, notably their more experienced colleagues.

Stakeholder involvement in decision making

Interdisciplinary projects will, to lesser and greater extents, involve interaction with a diverse range of stakeholders. Inviting stakeholders to be part of the decision-making process from the very start of projects, or as early as practicably possible, can help to ensure that the resulting decision(s) will be accepted (Renn *et al.*, 1997). This involves mediation among the stakeholder groups, usually through face-to-face discussions, a process known as cooperative discourse. Who should be invited to participate in the decision-making process and when tends to be determined by the type of project (private or public), its size and perceived impact on the environment, the time available for consultation and the attitude of the major stakeholders toward wider stakeholder involvement.

A systematic approach is necessary. The first stage is to identify the stakeholder groups that may wish to participate in the decision-making process, which is not always an easy task. The selection of professionals (addressed in Chapter 7) is related to the technical and managerial skills needed to complete the project. Assuming that the project is well defined and managed, this should be a relatively straightforward task. Similarly, the identification of specialist user groups should be relatively straightforward, if time-consuming. Identification of citizen groups with an interest in the project may be more challenging, especially groups that form to try and resist a new development. The group's purpose and identity will not be known until the project is made public and are likely to change over time.

Once the stakeholder groups are identified it is necessary to discuss concerns and values to see how these impact on the project (then determine the amount of decision making allowed).

Arnstein (1969) proposed a 'ladder of citizen participation' in relation to urban planning in the United States. Non-participation was represented by the first couple of steps on the ladder, moving up through a degree of tokenism to the top rung, citizen control. The further up the ladder the greater the decision-making power of the participant. Care should be taken in applying the ladder out of context, although the rungs of the ladder are useful in helping to illustrate the difference between being manipulated and having a large amount of control, and hence power over decision making. Within a project context it should come as no surprise that the large number of stakeholders will be allocated (by design or by accident) different levels of participation. This will be influenced by the project context, for example a public financed project or one funded from private funds, the size and complexity of the proposed building or structure and its location (see Chapter 6).

Participation is promoted as a positive way of making decisions, although critics argue that there is a danger of people taking more risks than if they had acted individually. In an AEC project environment there is also the practical challenge of whom to invite to participate. In recent years it has become popular (once again) to promote user involvement in the early design stages of projects, and although this is an admirable philosophy it does raise a number of practical concerns. For example, to represent users with disabilities requires, as a minimum, input from representatives with visual impairment, hearing impairment, mental impairment and physical disabilities. This may not always be possible in all situations and if only one person represents all disabilities then it is highly likely that the input will be skewed.

Within a project environment it is advisable to consider the responsibility of all of the participants, as some participants may have no financial stake or contractual responsibility for the decisions made, but may still want to have their voices heard through a participative process. This is why participants are often allocated different levels of power for their decisions, leading some to conclude that they have been manipulated and have not really had the opportunity to participate.

Group decision making

As a general observation the literature on group performance (Hare, 1976) and multidisciplinary teams (Ysseldyke *et al.*, 1982) tends to suggest that decisions made by groups are more workable and accurate than those made by an individual. Because of their broad range of specialisms multidisciplinary teams have been found to consider a wide range of potential solutions and groups have also been found to make more accurate, workable and rational decisions. Early research by Stroop (1932) found that group interaction produces a higher degree of creativity in relation to problem solving than an individual, although others subsequently noted exceptions to this observation. Lamm and Trommsdorff's (1973) and Taylor *et al.*'s (1958) research on idea generation through brainstorming exercises found that individuals outperform the group by a factor of 2:1. Furthermore, the individuals' ideas were found to be more creative than those of the group. The main finding from their research was that group pressures inhibit participation by some members. Individuals were found to participate less in small groups when they perceived their skills to be inferior to other group members (Collaros and Anderson, 1969). Emmitt and Gorse (2007) found that groups of four, five and six construction project management students consistently under-performed in brainstorming activities compared with the combined results of individuals working separately. Turn-taking, turn-blocking, fear of embarrassment and loss of concentration all had a detrimental effect on the group's ability to generate ideas. As a general finding each individual would produce fewer ideas than the group, but the combined output of the individuals out-numbered that of the group. Although the individuals were more productive than the groups they could not benefit from the evaluation of ideas within the group setting, which may be a disadvantage. Belbin (1981, 1993, 2000) found that groups of highly intelligent individuals often performed worse than a randomly formed group. This is because groups of highly intelligent people often have difficulty in agreeing on whose idea is the best.

Brown's (2000) review of group communication research indicates that the evaluation of ideas may be better dealt with in groups in which different perspectives can be used to analyse ideas. Although individuals and groups can both improve and limit idea generation and the evaluation of ideas, it is always important to consider the task and the attributes required from the individual or group. From the literature on idea generation and evaluation it is clear that for complex problems elements of the task should be broken up into individual and group exercises, combining the attributes of both individual and group problem solving.

Campbell (1968) reported that the contribution of more than one person increases the potential to solve complex problems quickly. Typically, in laboratory studies of human problem solving, subjects are presented with just sufficient data to solve the problem. However, a prime obstacle to solving 'real life' problems is selecting the relevant data from the body of superfluous, irrelevant and possibly misleading data. Campbell found that, generally, subjects take longer to solve problems when they have to differentiate between irrelevant and relevant data, something worth remembering when attempting to understand real life decisions. The results also show that some individuals appear to have the ability to instantly separate the relevant from the irrelevant, although when more than one person is working on the same problem the chance of finding appropriate information in a shorter time period is increased. In commercial environments participants must explore the various options and solutions available to them when solving problems.

Multidisciplinary teams have also been found to propose and consider a wider range of solutions to a problem when attempting to arrive at an overall solution (Ysseldyke *et al.*, 1982). Although Bales (1970) suggests that multidisciplinary teams may appear more productive in terms of alternative solutions generated during interaction, this could be a result of goal ambiguity. Yoshida *et al.* (1978) examined the content of multidisciplinary group interaction and classified it into contributing information; processing information; proposing alternatives; evaluating alternatives; and finalising decisions. They found that the frequency of individuals' participation, their perceptions and their contribution to multidisciplinary teams varied more than that in unidisciplinary teams. Stronger combined group forces often overruled individual expertise and experience. Thus, group consensus may go against expert opinion and information.

Contrary work by Littlepage and Silbiger (1992) found that, regardless of uneven and skewed participation rates, groups were able to recognise and use individual expertise confidently. Gameson (1992) found that construction specialists would rely on their own knowledge rather than suggesting that the contribution of others could be useful. Lee (1997) found that, even when professionals experienced difficulties in solving a problem because of limitations in their knowledge, they tended to rely on their incomplete knowledge rather than consulting a specialist for help. Lee found that this was particularly prominent in male participants, although as women assumed higher-status positions they also became associated with the trait of not openly asking for assistance. Mabry (1985) found no difference in the level of questions used by groups composed predominantly of men or of women. So gender difference may not be an issue in the use of question asking and information seeking, although construction participants who hold key positions may be reluctant to seek help, request more information or ask questions that may be perceived as revealing gaps in their knowledge. High-status professionals may still use probing questions to contest the knowledge of others, query the nature of information and/or undermine other contributors, but may avoid openly asking for help.

Communication

Effective communication is necessary to enable individuals to share in the understanding of problems and discuss the various solutions (Gouran and Hirokawa, 1996). Hosking and Haslam's (1997) observations of business relationships found that informal conversations within organisations were an important process for understanding what were considered as 'taken-for-granted' statements; thus conversation was essential to overcome ambiguity. Background information, clues and what may be considered 'small talk' are important for building relationships and are thus crucial for developing an understanding of unfamiliar contexts. Being able to enquire further into subject matter without the fear of embarrassment, ridicule or risk of offending others is achieved firstly through interpersonal interaction, which helps to build relationships and hence establish contextual information to inform the decision-making process.

Hollingshead (1998) found that, when members of a group were tasked with a problem, members became specialists in some areas but not others, and all members came to expect each participant to access information in specific domains; thus each individual took responsibility for specific tasks. Specialisation reduces the cognitive load on the individual, while also providing the group with access to larger amounts of information and knowledge. Newcomers to the group must communicate to explicitly identify responsibility for gathering and processing specific information; making assumptions about responsibility for problem solving results in less effective teamwork and duplication of tasks. In TPOs it is necessary to know who is most knowledgeable and skilled in specific areas so that individuals can assume key roles in related tasks. When members freely interact and openly disclose information, other members gain access to, and clues about, a member's knowledge and skills.

Extreme views and risk taking

Stroop has argued that the grouping of knowledge and experience acts as a moderating influence to restrict extreme views. The group's regulatory forces (imposed using a combination of conflict and group norms) control unacceptable views that are presented to it (Wallace, 1987). Unacceptable views are moderated through reactions, argument and conflict (Stroop, 1932). Contrary to these findings, Rim's (1966) extensive research on group and individual risk taking found that group decisions were more risky than those of individuals. The subjects' level of risk taking was based on the lowest probability of success they would accept when engaging in a task.

Group interaction changes the behaviour of individuals. As a general rule individuals will accommodate greater risks within a group environment than they would on their own. Bemm *et al.* (1970) found that individuals within groups would take greater risks even if the consequences of the risk taking would affect them personally. However, when group members were informed that failures associated with risk taking would be openly discussed and disclosed to the group, there was a shift to lower-risk decision making. Fear of group or public humiliation appears to temper risk-taking behaviour. Although such findings have been compared with the 'real world' context, the findings remain limited to laboratory-type experiments. In real life it can be very difficult to identify the level of risk associated with a decision. Similarly, it can be a challenge to state with any certainty whether the distinction made between individuals and groups applies to all commercial decisions.

Norms and decision making

Norms formed through communication affect group interaction and decision making. Group rules and norms are habitual, forming a backdrop or structure against which decisions are made (Hackman, 1992). This and other studies have shown that norms can have positive and negative effects on the decision-making process (Janis, 1972; Senge, 1990), the most common negative characteristic being groupthink, which is discussed below. Group norms can encourage cohesion and agreement, suppress critical inquiry (Janis, 1972; Cline 1994), reduce political input and increase rational discussion (Senge, 1990). Looking more specifically at language Giles (1986) noted that communication behaviour reflects the norms of the situation; however, it is often the communication behaviour and language that is used to define, and subsequently redefine, the nature of the situation for the participants involved. Group norms are powerful and will affect the way that actors interact within the group; however, some individuals are able to exert such a strong influence that it changes the group norm.

Keyton (1999) has suggested that high-status members might be exempt from the norm expectations that others are expected to follow. Generally, it is assumed that if a member deviates from the group norms the other members will react in one of three ways (Hackman, 1992):

- Group members may try to correct the behaviour, normally through pressure outside the group environment (a form of informal diplomacy).
- If deviation persists, other group members may exert psychological pressure through communication within the group, placing the deviant in an 'out-of-group' position (which the majority of people find uncomfortable).
- Finally, if deviation presents an acceptable alternative to the group norm and the behaviour (stance) is maintained over time it can influence other group members to accommodate the alternative norm.

Brainstorming and creative approaches

There are many approaches taken to group decision making. Often this group activity takes place informally within a meeting or workshop, or simply through the daily interaction of work groups going about their tasks. Techniques for stimulating group decision making, such as brainstorming and creative clusters, are addressed here.

Brainstorming

Brainstorming was proposed by Osborn (1957) as a technique for group decision making, the proposition being that brainstorming helps groups to be more creative and innovative. Brainstorming, if undertaken correctly and sparingly, can help to motivate members and assist with group development. Unfortunately, it is common for the term 'brainstorming' to be used to loosely cover any meeting called to discuss an unresolved issue, i.e. the term is used incorrectly (the participants discuss, they do not brainstorm).

The principles of brainstorming are that group problem solving should be addressed in stages and that each stage should follow some simple rules with the emphasis on the quality of ideas. It is common for one individual to lead the brainstorming exercise and to act as a scribe to capture the ideas generated by the participants. This facilitation role is demanding, in that the facilitator must manage the process and the participants' level of involvement (some participants may need to be encouraged to contribute), which requires some sensitivity. The brainstorming session should follow the stages below:

- *Define and agree the objective of the exercise.* Ensure that everyone participating understands the objective of the exercise and agrees to the rules. The objective should be kept as simple as possible.
- *Idea generation.* Participants are asked to randomly generate ideas, which are written on a flip chart or wipe board. This should last for a short period of time, perhaps ten to fifteen minutes. Ideas must not be evaluated, criticised or praised (which includes negative or positive body language such as shaking or nodding of the head). All ideas are welcome, and none should be rejected, no matter how ludicrous or unconventional it might appear. Participants are usually encouraged by the facilitator to build on the ideas of others (hitch-hike) in an attempt to build up ideas.
- *Idea evaluation.* Each idea is evaluated by the group to see if it is worthy of further consideration. Depending on the number of ideas generated it may be useful for the facilitator to combine ideas under headings or themes. This should be done in discussion with the participants as part of the evaluation process. Once the validity of the ideas has been discussed it is possible to prioritise the ideas into a list or set of options.
- *Agree and implement appropriate ideas.* Agree whether any of the ideas are worth developing in more detail and agree the actions required, timescales and responsibilities. A clear and positive outcome helps to reassure participants that their contribution was worthwhile. At the end of the session the facilitator should seek feedback on the brainstorming exercise from the group and keep the group informed of progress.

Whether or not brainstorming is an effective technique is open to debate. The business texts tend to be enthusiastic, although the academic community is more sceptical. Some of the practical challenges with the technique are that participants may need to be trained to be able to brainstorm effectively (which takes time and resources); it is difficult to ensure that no self-censorship takes place (participants may be apprehensive about having their ideas evaluated by others); and the evaluation of ideas is usually subjective (and often based on quantity of ideas, not the quality of ideas). On a more positive note, brainstorming can be very effective, especially in situations in which a group has failed to solve a problem and is unable to progress. Brainstorming is also used as an icebreaker in facilitated workshops and by academics to help newly formed groups to develop.

Creative clusters

Another format in which actors can interact is a creative cluster. These are multidisciplinary groupings that are set up, as cross-functional groups or teams, with the sole objective of addressing a specific issue. This may be a technical problem, for example how to detail a building component to make it more sustainable, or social, for example how best to select specific members of a temporary project organisation. The intention is to bring individuals together from different backgrounds with different knowledge, skills, attitudes and values with the deliberate intention of stimulating some creative tension and (hopefully) some new ideas and insights.

Once the problem has been resolved (or the resources have been used up) the creative cluster disbands. Success of the creative cluster rests on the knowledge of its members and their ability to interact creatively. Creative clusters are expensive to put together and therefore tend to be used sparingly.

Recognising and avoiding groupthink

Groups develop norms and their members are expected to conform to these. When the group norms are so strong as to resist ideas from outside the group (which are seen as deviant) then the group is deemed to be engaged in groupthink (Janis, 1972), a disadvantageous factor of group decision making. When groupthink occurs there is a danger that the group will make decisions that are wrong because essential information from outside the group has been ignored.

Groupthink usually occurs because the group exerts pressure, both explicit (from the dominant group members) and implicit (in the way the group acts), on all members to conform to the majority view. The overriding tendency is to agree with the majority view and suppress deviant ideas. The impression is that the group is making unanimous decisions that are correct because group members are not prepared to openly challenge decisions and alternative viewpoints are not investigated. This can happen with disciplinary groupings and within groups contributing to the TPO. The challenge for managers is to be able to spot groupthink before the consequences become catastrophic. According to Janis (1972) groupthink happens when certain conditions are met, such as:

- the group is very cohesive
- the group is insulated from external information
- all options are not appraised systematically
- the group is under stress, i.e. it needs to make an urgent decision
- the group is dominated by a strong individual (group leader).

Project managers and line mangers within organisations can mitigate groupthink by providing a culture for open and inclusive discussion and debate. Furthermore, the appointment of strong (overpowering) leaders to groups should be avoided, or dealt with if a strong leader emerges by removing the individual from some of the decision-making activities. Groupthink can occur in collaborative projects, as in the following example involving a partnering contract. From the outside everything looked to be progressing well and the participants all reported good working relationships and positive experiences. However, it was not until the project started to experience difficulties that some of the members of the TPO voiced their concerns about the decision-making processes. The project was a pilot project to demonstrate the benefits of strategic project partnering and the group norm was to agree with the project leader (a dominant personality) to try and ensure project success; everyone wanted it to be a successful project. By choosing to agree with the project manager, who had made some major errors of judgement, and choosing to ignore some of the more critical research findings being reported at the time, the TPO had entered into groupthink.

Design changes

Contributors to the decision-making process will be concerned with the allocation and reallocation of previously agreed resources, and therefore it is not surprising that disagreements and conflict manifest during the course of a project when attempts are made to reallocate (scarce) resources. Perhaps the most obvious example concerns changes made to the design and the impact this may have, not only on the performance of the overall building, but also on the additional work required to implement the change. This usually involves input from several disciplines to exchange information, discuss and agree an appropriate alternative solution. This means that time is required for the disciplines involved to review the appropriate literature, exchange ideas and opinions and hence make an informed decision.

The term 'design change' tends to have several meanings in the AEC literature, ranging from the natural evolution of design during the conceptual phase, through to changes made during the development of the design into detailed production information, to changes made to contract information during the production phase. For the purposes of this book, the term 'design change' is taken to mean a change to design information (drawings, schedules and written specifications) that has been approved and signed off by the client, regardless of the stage of the project. This means that proposed changes to the approved design information will result in additional work for those involved, which might not always have been budgeted for, hence the potential for disagreement and conflict over resources. Given that a building is an inter-related set of parts, it is common for a design change to affect other parts of the assembly, thus care is needed not to compromise the integrity, quality and performance of the building as a whole.

Given the potential consequences of design changes it is crucial that the project schedule allows space for a systematic, consistent and transparent process to manage design change requests. Quality management protocols tend to follow a simple and pragmatic sequence that all project stakeholders need to follow. First, the source of the design change needs to be identified, be it from the client, designers or contractors. Second, the reason for the design change needs to be confirmed, for example because of problems with constructability, delivery or cost. One this has been done the next step is for the design change request to be evaluated by the project manager and the design manager. If the request is valid then it is passed onto the appropriate discipline(s) to consider and a timescale for a response is confirmed. It is the individual disciplines, often working concurrently, that are best able to consider the consequences of the change request. The outcome of this process is either to reject the request, and state the reasons for the rejection, or to accept the change. If the change request is accepted (or an alternative solution proposed) then the implications of the change need to be qualified in relation to the process and the product. For example, the change may have positive or negative implications for the cost of the work, the timescale and the quality of the product. Changes may also have implications for health and safety, constructability and durability, not to mention the environmental impact of the building. Once a decision has been made, the final act is to seek approval from the client before the change can be implemented. Failure to follow the project protocol on design changes may result in abortive work and unnecessary disputes.

Information and decision making

In AEC projects information is an important resource; it is impossible to build without drawings, schedules and written specifications. Despite major developments in information technologies, the coordination of information in AEC projects remains a formidable management challenge. Coordination of information involves communication, discussion, negotiation and agreement between the various contributors. Interaction is required to ensure that there are no discrepancies between the information packages, which helps to reduce uncertainty and waste through future rework. Conflict can occur when information, for whatever reason, is not accurate and/or not complete, which usually brings about a request for further work in the best cases and results in abortive work and design changes in the worst cases.

Procurement of accurate and timely information is a challenge. Information needs to be correct and available when needed by the user. For example, setting up construction sites requires careful planning and logistics to ensure productive and safe working. Information is required about the design and the quantity of materials and components to be used to enable the scheduling of deliveries to reduce storage and unnecessary handling of materials on site, thus helping to make the process as efficient as possible. Changes to the design may affect deliveries of materials to the site and affect the sequencing of works packages in some extreme cases; again this can lead to disagreements and conflict between contributors.

Practical challenges

The main practical challenges in an AEC context tend to relate to the amount of time available to make decisions. Some projects are not particularly well resourced given their size and complexity and this often leads to problems with coordination of information and the ability of the participants to make informed and considered decisions within a finite time frame. The way in which the project is managed can also have implications for decision making, for example not having enough meetings to allow participants to discuss issues face to face. Poor allocation of resources to tasks can also affect decision making, often giving some individuals too much time and others too little. These challenges can be overcome by carefully mapping the process before the project starts and asking the contributors for their input into the scheduling process.

By mapping the process it is possible to identify responsibilities for decisions and then communicate a decision-making protocol to the project participants. Many problems are created simply through not knowing who is responsible for specific decisions. Clear direction from the project manager, combined with transparent decision-making project protocols, can help to eliminate much of the uncertainty associated with decision making in a dynamic project environment.

As already highlighted, the decision-making process is coloured by the individuals contributing to the process. It follows that one of the prime responsibilities of the project manager is to assemble the most appropriate individuals and organisations for the project context. This involves selecting stakeholders with the appropriate technical skills and also with the ability to interact effectively with the other participants. Failure to interact and communicate will often lead to inappropriate decisions, which will have consequences for the building and also for the project stakeholders.

End of chapter exercises

- Early in the project you find that decisions are being taken by the main contributors without adequate consultation with the other disciplines. How do you address the problem?
- During a series of early meetings with the main project contributors and the client representative you suspect that no-one is prepared to challenge decisions. Your concern is that groupthink is starting to manifest. What actions could you take to prevent groupthink and promote informed decision making?
- A sub-contractor has made a change to the approved design information and implemented it on site without following the appropriate protocols. What should you do and why?

Further reading

Additional detail on decision making can be found in *A Primer on Decision Making: how decisions happen* by James G. March (1994).

6   Context

Management thinking, just like architectural style or the clothes we wear, is not immune from fashion, which fuels it. A characteristic of most management concepts is that they promise (often dramatic) performance improvements, while also leaving considerable scope for interpretation and implementation. Although this may be a clever ploy on behalf of those promoting a new way of going about our business, it can and does often lead to confusion. The tendency is for these management fads to become detached from their original meaning as they are interpreted and implemented by practitioners (Abrahamson, 1996). This means that it can be difficult for practitioners to evaluate new approaches in terms of how concepts and philosophies relate to their business and their portfolio of projects. It also means that transferring ideas from one sector to another can be problematic (Brensen *et al.*, 2005; Jørgensen and Emmitt, 2008). It should also be recognised that different organisations will find one approach more advantageous to their particular *modus operandi* than another, whereas another (similar) organisation may take a very different view.

According to Hartley (1997) we simply do not know enough about how humans behave or how groups perform to be confident in guaranteeing that one approach or technique will have the same effect on everyone. There are no set recipes for achieving successful projects; merely some approaches prove to be more effective than others for a specific client at a particular time and place. Indeed, to promote one approach as better than an alternative for all situations would be rather reckless. Instead, it is more important to understand the contextual drivers of projects before taking a decision about how best to manage those involved. Similarly, it is necessary to understand one's organisation rather than disconnect from intelligent and rational thinking and adopt the latest management trend, often with unforeseen consequences, simply because everyone else appears to be doing so. The ability to engage in critical thinking and go one's own way can give organisations a distinctive edge over their competitors.

Developing an understanding of the fundamental issues explored in the preceding chapters will help professionals to consider the context of their businesses, their clients and their organisational project portfolios. Similarly, the ability to quickly analyse and respond to the specific project context in a rational and considered manner may represent a significant competitive advantage to the organisation and also add value to clients' projects. Indeed, the approach taken to the management of AEC projects should be a direct response to a unique context. It is the response to context that ultimately brings about the final physical artifact. Even with repeat clients, repetitive building types and a relatively stable TPO, the physical and social contexts of the site will still differ from one project to the next. The purpose of this chapter is to explore the main contextual characteristics that, to lesser or greater extents, underpin, inform and shape AEC projects.

Fundamentals

The word 'context' comes from the Latin *contexere* (weaving together) and is used to describe the setting of an event or building: the weaving together of people and physical artifacts. Context is a crucial determinant of project success, as is the choice of appropriate processes, people, methods and tools. Context also influences the transfer of ideas and technologies and hence the rate of adoption or rejection of new ideas and methods.

Buildings and civil engineering projects are designed for a specific purpose and are constructed in a unique environment using a variety of materials and components over a defined timescale, to be functional for a designated period of time. In essence an AEC project is the harmonious weaving together of people, materials and place. The ability to achieve synergy between people, materials and place throughout the project is founded on the ability to analyse the project context. It is necessary to explore a number of potential approaches before deciding on the most appropriate 'fit' for the client and the organisations contributing to the project. This means that we need to understand the project's contextual characteristics before making decisions about the procurement route and contracts to use, or before we start to put the TPO together. The manner in which the TPO engages with the project context will impact on the success of the process and also the performance of the product. The most dominant factors in determining context relate to the client, the physical location of the site, the social context and the time in which the project is set, which are discussed below.

An evolving context

Given the long timescales associated with AEC projects from inception to completion it is necessary to recognise that the context is changing during the life of a project. Most obvious are the physical changes that happen on the site, as these are visible and can be quite rapid in the case of demolition and the erection of prefabricated elements, but other, more subtle, changes occur as organisations and individuals enter and leave the TPO. Time also has a role to play, with local and global events (such as localised flooding, a major shift in the economy affecting interest rates and confidence levels, and global terrorist attacks) influencing the decisions made within projects. Client context might also change, especially on projects with a long duration, as client organisations are dynamic, and it is quite probable that the main client contact person will change because he or she is promoted within the organisation, or may move to another employer. Contributing organisations might also evolve, growing, shrinking or even going out of business, bringing about changes in the composition of the TPO as the individual contributors are replaced.

Because the context is changing over the life cycle of the project it is necessary to reconsider the context at regular intervals to see how this impacts on the TPO. This can be built into, or linked to, regular review processes.

Responding to context

Each project should be tailored to suit the individual requirements of the client and the context of the construction site, which may vary significantly between consecutive projects. Projects are characterised by the values of the client and the TPO. They are also characterised by the values associated with the construction site (contextual values) – thus transferring the same client and TPO to a new site will create a new context and associated values. The client, site and social context will underpin the managerial context, all of which are discussed in more detail below.

Client context

AEC projects involve an interplay between two complex and highly dynamic systems: the client organisation and the participants of the TPO. The term 'client' is commonly used in AEC to describe the person(s) or the organisation(s) funding a project. Alternatively, terms such as 'building sponsor' and 'customer' may be used to describe the same thing. It is the client, be it an individual or a large corporation, that funds the project, usually by borrowing money. Therefore a number of other parties, such as banks, private equity investors, pension funds and insurance companies, will have a financial stake in the project. In some projects these stakeholders may participate directly, with investors represented at strategic project meetings. Alternatively, the investors can indirectly influence the project by laying down a number of performance criteria to be met, for example stringent security measures, which will be incorporated into the strategic and project briefing documents. Understanding how each project is funded and mapping the financial stakeholders may be beneficial in helping to identify some of the project values, although clients are often reluctant to disclose such commercially sensitive information.

On small domestic work the client may be the wife or husband, representing the interests of their family unit. In this situation the client is committing their own finances and emotional capital to the project. Capture of the family's needs and aspirations can be undertaken on a face-to-face basis with all family members. On larger residential developments and other projects, such as commercial schemes, the client is a business organisation. In this case the client's representative may be one or more people, charged with doing a job. Rarely are they themselves the investors, owners and users of the building. This makes the capture and communication of 'client values' particularly difficult to achieve in practice and necessitates representation of, for example, user groups in the decision-making process. Although client briefing (architectural programming) is outside the scope of this book, it is necessary to consider the level of client experience, their commitment to the project and their willingness and ability to participate in the decision-making process.

Client expertise

Clients may be defined in relation to their previous experience of construction (Emmitt, 2007):

- *First-time clients*. The first-time client will need guiding through the design and construction process. This client may only commission design and construction services once, for example a house owner wishing to provide more space for a growing family. Here the brief will be a bespoke document. Effort may be required to ensure that the client fully understands the implications of the decisions being made. Visualisation techniques and simple graphics are very helpful in this regard.
- *Occasional clients*. The term 'occasional client' tends to be used to describe people or organisations that commission design and construction services on a relatively infrequent basis. The gap between commissions may be lengthy and each type of project may be very different to the last. Thus learning from previous project experience may be challenging. The brief is likely to be a bespoke document.
- *Repeat clients*. Repeat clients tend to be major institutions, businesses and organisations with a large property portfolio. Typical repeat clients would be food retail businesses, hotel chains, etc. Here the commissioning of buildings is more likely to be part of a strategic procurement strategy, closely linked to the business objective of the organisation and its facility/asset management strategies. There may be an opportunity to establish integrated teams that move and learn from one project to the next. Similarly, the ability to make improvements to how design and construction activities are managed is also present with repeat clients. With repeat clients the brief may include elements common to previous projects, which represent the values and knowledge of the client organisation, codified in a standard brief.

Client types

Professional advisors need to represent the interests of three distinct groups: the building owner (client owner), the building users (client user) and society (client society). These three groups value different things at different times in the life of the building. The predominant focus is on when the building is completed and taken into use, when durability, usefulness and beauty may be used as an expression of the primary view of each of the three groups respectively. But there also exists the perspective of the value of the building in the future or for future users, and the value while the building is being realised. Value cannot be measured or expressed and communicated explicitly, but must be learned and understood through a process of interaction and exchange. This transaction is most evident in the briefing process, a learning process for all participants.

- *Client owner.* The term 'client owner' is used to describe a person or organisation responsible for paying for the project. Clients may have short-term (developer) or long-term (owner) interests in the completed building. Financial commitment makes the client owner unique because client users and client society rarely make a direct financial contribution to the construction project.
- *Client user.* User involvement is regarded as a key element of successful projects. Many client users, for example office workers, will have a self-interest in the building project, but they will not have a financial stake in it. With regard to publicly funded buildings, such as healthcare buildings, users will have an indirect financial stake through the taxes they pay to the state. The client user will be influenced by the purpose of the building and its physical location. Building users can make a significant input to the data collection exercise as it is they, rather than the building designers, who interact with specific areas on a daily basis. Soliciting their views and listening to their requirements is a fundamental part of the briefing process. Identifying users can be problematic for many building types as users often constitute a disparate mix of people and groups. It is often impractical to try and capture the views of all potential users; instead representatives of specific user groups are identified and brought into the project briefing phase. For example, personnel managers, facilities managers and building maintenance managers all have a part to play in representing the interests of a wide cross-section of building users. User consultation processes, for example questionnaire surveys and workshops, help to provide essential information and knowledge for analysis, the results of which can be taken forward as a set of value parameters. This is 'first generation' user involvement and some flexibility (and vision) needs to be considered so that the second and subsequent generations of building users are also considered in the development of the brief.

- *Client society.* The term 'client society' is used to refer to stakeholders who may have an involvement in the project but are not part of any contractual agreement. Neighbours, local interest and pressure groups, town planners and building control officers will all be stakeholders, although in many cases these individuals may never use the building. Small extensions to residential properties may be of little concern to the community at large, but may have a significant impact on the immediate neighbours. For larger and more prominent developments there is often a need to engage in some form of public consultation exercise. These may be linked to the town planning process, or may be initiated by the building sponsor in a proactive attempt to hear the opinion of the local community. Public consultation exercises need to be carefully organised to allow members of the community a chance to contribute in a positive and timely manner. Similar to user involvement there must be consideration of future societal needs as the building will still be around long after the current members of the client society.

Client interaction

The parameters that help to determine the quality of the finished product are time, money and value (and risk). Time is usually a significant factor in the composition of the TPO. Start and finish dates may influence the people who are available and the overall programme can exert considerable pressure on the TPO. The cost of the consultants may not be a significant concern compared with the overall initial cost of the project, but the cost of labour and materials will be. Although there is often a trade-off between these three parameters, attention should also be given to the level of interaction between the three client groups and the TPO.

Success of the project is dependent largely upon the client owner's perception of the service provided by the temporary project organisation and the fitness for purpose of the finished building. Success is concerned with the difference between expectation and delivery. It is about delivering value to client and building users alike.

Strategic briefing

The values of the client need to be explored and defined through a well-managed briefing process. The values of the project team are defined by the way in which the team is assembled (individual values and competences), the procurement route used (which influences the attitude of actors) and the way in which the team is managed (which influences interaction). As already noted, the values associated with the physical location of the building (or more correctly locations, as much of the work is carried out off-site) are influenced by a multitude of factors determined by the site's physicality and social interactions, which is explored in more detail below.

One of the earliest tasks in the project briefing process is the establishment of the strategic brief for the project. This involves making some very important decisions relating to the project, a process sometimes referred to as optioneering, in which the client and key advisors choose from a number of options. Here the (economic) feasibility of a number of options is reviewed before the project proceeds any further.

Physical context

In building design a lot of emphasis is, not surprisingly, placed on the site: the genius of the place (*genius loci*). For many of us our relationship with the planet, our connectivity to the earth, is mainly through the built environment, which is perceived and experienced through our senses. We work, play and sleep in buildings and interact with the built, artificial environment on a constant basis. In architecture, the word 'context' is used specifically to refer to the physical (built) environment in which the building project is to be placed. We design by considering something in relation to its larger context, for example a handle on a window, a window in a wall, a wall in a building. We need to consider how individual components are woven together to achieve the required building performance.

In contrast to manufacturing, AEC projects are influenced by the physical characteristics of the site on which the building will rest, i.e. the new building will need to weave into the existing fabric. The physical context of the site will (or should) define the uniqueness of the design, the realisation process and hence the final product. The physical context may also influence the choice of participants contributing to the project, a point explored in more detail later.

Physicality

Every site is defined by its physical, three-dimensional context. Each will have its own particular ground conditions, its own micro-climate, its unique juxtaposition with neighbouring buildings, roads and boundaries. Sites will also have specific design constraints, such as access points, availability and proximity of services, existing levels, trees and hedgerows. Sites also have a memory, a trace of previous uses and associated values, which might be an important consideration when inserting a new building into the landscape or undertaking repair and restoration projects. Sensitivity to the local environment, the pattern or grain of an area, combined with the physical characteristics of a specific site serve as design generators, with some taking greater priority than others in the design process.

Construction is a process of transition. Sites have a unique identity at the start of a project and a new, equally unique, identity when the project is complete. This new identity may be subtly different from the old in the case of sensitive refurbishment and repair projects, or drastically different in the case of demolition, identified by material absence for a comparatively short period and replacement with a new artifact for many, as yet undetermined, years. Indeed, it is sometimes the shock of the new that influences our initial reaction to a new building in an existing streetscape. Insertions into the townscape influence the grain of a town or city, and not surprisingly there are many stakeholders (client society) who may wish to influence the proposed development, by opposing or supporting it.

Following this line of reasoning it is possible to see the physical site as having a value and a number of values associated with it:

- *Site value*. This is represented in financial and cultural terms.
- *Site values*. These are represented by individuals' memories of buildings and events and their perceived meaning.

Development of the site will change the value of the site and will also change the values associated with it.

Physicality and social interaction

The physicality of the site will set the stage for social interaction, helping to determine some of the project's stakeholders. This will include the legal owner(s) of the land, who may or may not be the project sponsor, and their legal representatives. Selection of the temporary project organisation may also be influenced by the geographical location of the site, for example the appointment of an architect situated within the locale and the use of local suppliers and contractors. Thus local knowledge can be utilised while at the same time helping to stimulate the local economy and possibly reduce the environmental impact of the project by reducing, for example, transportation costs. Immediate neighbours will, quite rightly, have an interest in any development that might affect them, and they will express their views directly to the client or client's project manager, and indirectly through the town planning authorities. Similarly, local user groups and pressure groups are likely to take an interest, either in supporting or resisting the proposal(s) for the site.

Location of the site will also determine which local and district councils are 'responsible' for that site, that is, whose jurisdiction the site falls under. Most obvious are the town planning department and the building control department, but other departments such as environmental health, refuse, highways, drainage, police and the local fire department will all become stakeholders in the project (if only for a short period of time), represented by one or more representatives of that department. Bylaws differ from town to town and the interpretation and application of legislation and guidance documents may well vary from one region to another.

The extent to which these disparate participants are included in the design process, and hence the project, will depend on the amount of inclusivity permitted or encouraged by the project team. Some clients and designers are happy to work with neighbours and the relevant authorities to help develop the design and acquire the necessary approvals as quickly as possible with as little conflict as possible. Others may be less keen to have their ideas compromised by officials and will adopt a less inclusive, often confrontational stance. It is the attitude towards the various stakeholders that will determine the membership status of these individuals, as fully participating members of the project or merely occasional visitors to a small number of meetings. This assortment of individuals, groups, departments and organisations will all have a stake in the project, but it is important to recognise that they will not be party to any formal contracts relating to the project *per se*. Some of these may be reluctant stakeholders, involved simply because a neighbour has embarked on a new development that is perceived to be problematic.

Research into the specification of building materials and components has found that building control officers and town planning officers can have a significant impact on detailed design decisions (Emmitt and Yeomans, 2008). A feature of collaborative/inclusive design is that these individuals are included in the design process, rather than simply consulted. This does not mean that professionals should take all legislation at face value. Sometimes it may be necessary to challenge legislation and demonstrate a more appropriate approach for a given context, for example in the case of restoration to a protected building which may require a design solution that is not aligned with current practice and legislation, but that is suitable for the building.

Site selection

For some clients there is no choice about the site to be developed; it is a fixed project parameter. However, many clients will review a selection of sites, for example clients engaged in commercial property development and clients expanding a large property portfolio such as food retailers. Site selection will be based on a number of value criteria relating to the client's business and will form part of the strategic briefing process. As noted above, the choice of site will set in train a number of interactions with local community groups and local council officers dealing with aspects relating to town planning, highways, fire, policing and building control.

Social context

Any social situation is a sort of reality agreed upon by those participating in it, or more exactly those who define the situation. This is known as social constructionism, which relates to the way in which individuals make meaning from their social situation. Everyone who enters a situation does so with preconceived definitions of what is expected of him/her and the other participants. Such beliefs, which include expected interactions, are established from experience of previous groups to which the individuals have belonged.

Thus, each situation confronts the participant with specific expectations and demands. Such circumstances generally work because most of the time our perceptions and expectations of important situations coincide approximately. Culture, society and individual power play an important part in the norms that govern the way that social groups act. The same aspects will also influence the behaviour of professional groups, but groups may also draw on those members with nominated roles and perceived expertise, skill and experience to determine who is allowed to interact, make decisions and play lead roles. However, some people, regardless of professional skill or experience, can use communication techniques to exert influence, enabling them to gain power over elements of group behaviour and decision making.

Understanding buildings (architecture) is not just about materials and technologies; it also concerns the social context in which they are conceived, delivered and used. Social context is determined by the time and place in which we live.

When analysing the social context of a project the time allocated to a project is, perhaps, the most obvious factor, but there are a number of other time-sensitive factors that can affect the project, which are summarised below.

Legislative framework

Legislation (comprising regulations, codes and standards) and the legal constraints pertaining to the site (such as rights of access and rights of light) are design generators.

Legislation changes over time and may change during the course of a large and complex project, leading to design changes and in extreme cases the involvement of additional specialists to deal with the changes. Some of the more recent changes relate to more stringent health and safety legislation, greater attention to anti-terrorism measures in and around buildings and incremental changes to the building regulations (especially the thermal performance of buildings).

Socio-economic issues

Local, national and international economic fortune will influence the project in a variety of ways. In a very busy period it may be impossible to get the organisations and or people best suited for a particular project. Suppliers of materials may be working on long lead times and so designs may need to include materials that are easier to procure. The choice is to wait, or more commonly to proceed but with perhaps less-suited contributors. In a recession, the choice of organisations and individuals is greater and the competition for work will be more intense.

Architectural and managerial fashion

Architectural fashion can sometimes influence the type of procurement routes used and hence the make-up of the TPO. For example, the use of off-site manufacturing and prefabrication can, in some cases, negate the need for a general contractor. Preassembled units are delivered to site and craned into position. The foundations, services and associated tasks can be undertaken by specialist sub-contractors. Similarly, the choice of a topical management innovation can have implications for the performance of the project team, both positive and negative.

Complexity

The complexity, diversity and sophistication of the society in which the site and project is set can have a major influence on the extent to which individuals participate in projects and also on the extent of legislative and political constraints. Obviously, this varies around the globe, from state to state, from region to region and from city to city. Sensitivity to different layers of complexity, sophistication and diversity should be addressed when mapping the project, helping to inform the planning of the project process.

Cultural differences

Different organisational cultures and individual cultural differences will be present within the temporary project organisation. Differences between organisational cultures tend to be revealed at the interfaces between the organisations. Individual cultural differences, the values of the contributors and the values and traditions of the society in which the project is set, may be less obvious. Working across international borders will need a heightened awareness of cultural differences and their potential impact on the way in which the project is organised and realised.

Political factors

At the national level changes in government and the effect of global money supply can influence the way in which projects are financed and procured. On a local level the political make-up of the local council may influence the type of development supported. An understanding of the political drivers relating to a particular site/location can influence the type of development and the type of procurement route used.

Leadership context

Once the context has been researched, analysed and understood it is then possible to think about appointing the managers (or project managers), choosing the most appropriate management framework and assembling the TPO best suited to the project parameters (see Chapter 7). Managers should set the scene for effective and efficient interdisciplinary working, i.e. create the most appropriate working conditions, from the very start of projects. This includes the ability to encourage active engagement, participation and empowerment.

Selection of an appropriate manager for the project context is important. It would not be sensible to take a project manager who had an excellent record of dealing with highly complex fast-track commercial projects and assign that individual to a sensitive refurbishment project; he or she would be unfamiliar with the environment and would in all probability not be equipped with the necessary knowledge for the new context. A great deal of effort and emotional energy would be spent on familiarising him- or herself with the characteristics of the new building typology, something that takes most professionals many years to acquire through experience of working on similar types of project. Although this may be an extreme example, project managers sometimes find themselves working on projects that are different to their previous assignments and their individual knowledge and skills may not be well suited to that project context, which can often result in inefficient processes and waste. Given sufficient time and resources, it may be useful to engage professionals with experience from different building types to bring fresh thinking into a project through knowledge transfer, although this is not recommended if resources and timescales are constrained.

Leadership

Leadership is an emotive subject. For some authors it is the inter-relationship of a client's needs and the restrictions of the site that 'ensures' the position of building team leader to the architect, but recent trends have seen the leadership role pass to other management-orientated professions. Leadership is important because it is the most active link with the building sponsor, an important link if business opportunities are to be maximised. Leadership is also important in terms of delivering value through design, and consideration must be given to contractual arrangements that allow architects to control design quality, rather than delegate it to others. Leadership skills are particularly important in developing an effective communication culture throughout the project life. Projects represent a temporary overlap of authority and there may be rivalry of power within the TPO, which is associated with the allocation of resources. People tend to remain loyal to their employer, not the project, which can cause difficulties for the project manager.

Managers should have the experience, competence, ability and energy to:

- provide appropriate, clear and consistent leadership
- develop and maintain project values

- compose and hold disparate groups together
- develop empathy and establish appropriate extent of trust with all contributors
- communicate effectively within and between different levels
- design effective project communication structures
- encourage interorganisational and intergroup communication
- implement systems and tools that enable participants to collaborate effectively
- arrange and chair meetings
- develop relationships with informal leaders and organisational gatekeepers
- map and facilitate value chain activities
- establish and develop the project attitude to risk and innovation
- communicate and reinforce project goals, safety culture and quality standards
- benchmark project performance
- deal with crises quickly and openly
- manage conflict to the benefit of the project
- maintain an ethical approach
- maintain and develop the project culture (avoid entropy)
- provide regular opportunities for feedback and learning.

Case study

Two project managers working for the same project management organisation were monitored over a twelve-month period. The project managers were working on 'identical' fast-track health-care projects. Both buildings had been designed by the same design team to perform the same function and so the buildings shared a common set of construction information. The buildings were located adjacent to one another on a large, flat site that was owned by the client. Both buildings were started at the same time and had an identical programme. Each building was to be constructed by a different prime contractor, who had been selected through a negotiation process. Both contractors had worked for this client before and were familiar with fast-track health-care buildings. Although the contractors used their own preferred sub-contractors some of the sub-contracted packages were common to each project, such as the steel and concrete and internal fit-out. This meant that each project manager was charged with managing different TPOs.

During the monitoring period both project managers demonstrated exemplary skills: both projects were completed on time (to the revised completion date, which had been brought forward when the projects had been accelerated in response to the client's needs) and to budget (revised, higher budget because of acceleration of the work). There were no reported major accidents on either site and the build quality of the buildings was very high. Comparing the two project managers, similar age, similar educational background, same sex, there appeared to be nothing different in their performances. Both were rated equally highly by their employer, and were considered the best project managers in their organisation. So, we could conclude that they have equal capabilities, but we would be wrong.

Shortly after the completion of these projects both project managers were assigned new projects. Once again, these were fast-track, highly complex commercial projects, with many similarities to those that they had just completed. The first project manager was able to assemble a core team that included almost everyone from the previous project (one member was new to the team). The second project manager found that new members had to be brought into the team because those contributing to the previous project asked to be put on other projects. So, why the difference?

The difference between the two project managers related to their interpersonal abilities. The first project manager took time to visit the construction site and to talk informally to the project team, albeit briefly. When a problem arose he solved it by telephone and/or by visiting the site to talk it over with the team before making a decision. In doing so he had established a rapport with the team members; he placed considerable value on face-to-face contact. The second project manager appeared to be less comfortable with face-to-face contact, preferring to manage the team remotely, rarely engaging in informal conversations when visiting the construction site. When a problem arose his tendency was to solve it by sending emails and referring the team to the conditions of the contract. The first project manager was visible and approachable, whereas the second project manager was less so. It was this difference in personality traits that distinguished the two project managers and appeared to make the first project manager's job easier than that of the second (who had to engage in team building from scratch).

Process management context

In an ideal world management approaches and process models should be implemented in response to the project context. This means selecting the most appropriate people, implementing the most appropriate tools and applying the most appropriate processes (Figure 6.1).

People factors have already been discussed throughout this book. How individuals interact, communicate and are able to work together is an essential consideration for all projects. Tools and technologies have also been addressed in the early chapters, with the argument made for synergy between people and the tools that they use. In this section the focus is on process factors which stimulate social interaction and effective working within the TPO.

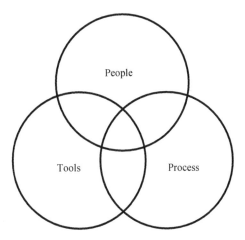

Figure 6.1 Process management context.

Process management

The majority of organisations and managers tend to favour one or two approaches that have been found to work well on previous projects. Standard process models, such as the RIBA Plan of Work, are often adapted, based on experience and to suit the organisation and its client/project portfolio. The result is a bespoke approach. This can create a challenge when two or more organisations interact, because each is familiar with their interpretation and application, which may differ slightly between organisations. Early discussion about the tools being used can help to eliminate misunderstanding later in the project.

Process management is usually divided between the creative design phase and the more practical construction phase. Separating the two functions partly reflects the differences between managing design and managing construction, as well as reflecting the historic separation between design and construction activities in the AEC sector. Arguments for better integration of design and construction abound, and, although a worthy argument, the differences in management approaches in the two stages need to be recognised.

Design activities by their very nature are concerned with identifying problems, problem framing, and proposing a variety of solutions. This has led to the development of design management (architectural design management) literature (Emmitt, 1999) as something distinct from the construction management literature. Creative design phases require more space for exploration of possibilities and preferences than the physical realisation phase. The challenge for managers is identifying the interfaces between these two activities, which may be clear-cut in some traditional forms of procurement, and applying an appropriate managerial framework.

Two management approaches that have been widely promoted in recent years are project partnering and lean construction. Both are concerned with process improvements and in their own way aim to improve integration between interdisciplinary contributors. The principle of project partnering is based on cooperative relationships and long-term relationships (in the case of strategic partnering). Lean thinking has been taken from the manufacturing sector and applied to the AEC sector, mainly under the guise of 'lean construction'. The philosophy is to reduce waste and maximise value for the end customer by continual improvement. This approach also requires cooperative partnerships and a change in attitude to how projects are procured and delivered.

Practical challenges

As noted elsewhere in this book it would appear that one of the most common ailments to afflict projects is starting the project without due consideration for the underlying contextual issues. Although this may appear a rather foolish thing to do, the root cause is usually related to a client's desire to start the project as quickly as possible. The timescale and fee level may also contribute to a collective desire to rush into the project, with contributors keen to get the project moving and hence generate some income. The result is a responsive, emergent approach with the project developing in an organic manner rather than from a sound base.

Ignorance of context (e.g. being surprised by the reaction of the local community to a new development) can be detrimental to the harmonious development of a project. It is necessary to do some research into the factors identified above before starting the project. This is particularly important when working on sites geographically distant from one's own (familiar) environment. A simple three-step approach may help:

- *Step 1.* Review and analyse the project context.
- *Step 2.* Review and prioritise project values.
- *Step 3.* Review process models for an appropriate fit.

Once this is complete it is then possible to consider the composition of the best TPO capable of adding value, i.e. the people and the organisations. This is discussed in the next chapter.

End of chapter exercises

- You have successfully completed a project for a major client. The client has asked if you will manage a new project and is insisting that the same TPO be used. The building type is different from the previous project and the site context offers very different challenges from the earlier project. What should you do, stick with the same (familiar) TPO (option A) or argue for a new (unfamiliar) TPO that is better suited to the context (option B)? List five advantages and five disadvantages for each option.

- You have been shortlisted by a client to work on a major project. One of the contract conditions is that all contributors must work with the client's bespoke project delivery system (an integrated design and construct process model). The model is unfamiliar to you and your organisation. What should you do?

Further reading

Context is addressed in the majority of architectural publications. For further insights into the client context see *Understanding the Construction Client* by Boyd and Chinyio (2006), and for further guidance on client briefing see Blyth and Worthington's *Managing the Brief for Better Design* (2010).

7 Assembly

Once a thorough understanding of the project context has been established and discussed with the client it is then possible to consider the assembly of the TPO. This involves assembling the most appropriate organisations and individuals to realise the project. These may not necessarily be the most famous or best organisations (as considered by their peers), but those most suited to the specific project context. Consideration should be given to compatibility of the project values and the values of the potential contributors. It should not come as any surprise to state that this is an important task that will influence the performance, profitability and outcome of the project. Get it right, and few people will notice, but get it wrong and everyone (with the benefit of hindsight) will be quick to point out the incompatibilities and the problems created. What may come as a surprise to some readers is the reality that this aspect of projects is not always given the attention it deserves. Too often clients are in a hurry to start the project and the TPO is assembled more by circumstance than by design. This may not necessarily be unique to the AEC sector, but it appears to be a strong characteristic of AEC projects and an issue that many professionals are concerned about. As highlighted in the previous chapter, in an ideal world the assembly of the TPO should be started before the project briefing stage commences, although this is not always possible.

Neither design or construction quality can be achieved merely through the setting and achievement of technical specifications. Quality is also influenced by the way in which people act and interact, creating and solving problems as the work proceeds, and the manner in which managerial and quality controls are applied. Achieving high quality involves the use of the most appropriate systems and the most appropriate mix of organisations and people for the project context. As highlighted earlier in this book, the manner in which the various actors interact will create and shape the project culture. The design of this dynamic and constantly changing social system should not be left to chance, although it may surprise some readers how little attention this aspect of project management is given in practice. Too often the team is allowed to grow organically; sometimes things work out well, but more often difficulties with communication and misunderstanding manifest because of differences in values. This incompatibility should have been tackled earlier in the process. Social engineering of the TPO is an important task for the project manager. Selecting the team members and massaging their egos to the collective benefit of the project takes a lot of skill, experience and sensitivity. Technical incompatibility may be relatively easy to avoid, but personal chemistry and compatibility is more of a challenge.

In the vast majority of construction projects the participants are brought together to work on one project only. Following completion of the project, or more accurately completion of a participant's work package, the relationship between the individual and the project stops. This means that, with the exception of large and repetitive projects, it is common for the TPO to be composed of different actors from the previous one. This is often true even when the same organisations are involved, simply because different individuals within the organisation have been assigned to the project according to internal workload commitments. Thus communication is required to support industrial relations and hence provide the means by which teams can develop quickly and effectively. Relationships can be volatile and adversarial, making it difficult to form and thereafter maintain interorganisational relationships. To a certain extent initiatives such as partnering, strategic alliances and integrated supply chain management help to mitigate the affects of fragmentation, although, as with more traditional approaches, there is still a heavy reliance on the ability of the team members to interact and communicate effectively. Thus, regardless of the approach adopted, the basic tenents of running a project remain the same, i.e. we are reliant on getting the right people together for the right job. Competences and the development of competent practices are a key factor in the success of the TPO. The attributes and actions of key personnel strongly influence the success or failure of the project.

Fundamentals

It is in the early stages of a project that the majority of opportunities are generated (or lost) and the risks minimised (or generated). Early decisions also influence the health and safety culture, the attitude to quality, and the social and economic conditions that are subsequently infused within the project. Studies have shown that investing in early team building can be beneficial to the smooth running and successful completion of projects, with the time and effort invested up front recouped as the project progresses to completion. The establishment of the TPO is an area most associated with the project manager, although clients and design managers also have a role to play. Building the project's system architecture is a highly complex undertaking, requiring an understanding of interpersonal communication traits and an appreciation of how groups and teams function effectively. Research on professional interaction in live construction projects has found that communication does not always follow formal channels. Professionals pursue social structures that benefit the individual organisation, i.e. they will use informal communication routes to achieve their objectives.

Architects, engineers, project managers and other key consultants must demonstrate the value to their clients of starting projects from an informed, considered and solid foundation. Far too often projects are rushed into without adequate understanding of the importance of the early phases. Research in business management has consistently revealed weaknesses in the front end of poorly performing projects. This can be found in AEC projects, with problems encountered in the realisation and use phases traced back to poor decisions early in the life of the project. The recipes for successful projects appear to be related to the assembly of the most appropriate participants and comprehensive briefing to determine project parameters. Here the creation, retention and realisation of the design vision throughout the life of a project are of paramount concern. Building sponsors must accept that too much haste in the early days of the project life may have severe consequences.

A strategic approach

Research findings and anecdotal evidence from practitioners suggest a strong correlation between successful projects and the time spent assembling the most appropriate people and organisations to work together. Time invested early in the life of a project can make a significant impact on the future ability of the actors to interact efficiently and effectively. In some respects it is common sense, yet far too often projects are conceived and launched without pausing to think of the consequences. Sponsors of building projects may be reluctant to invest resources (money and time) in the preliminary composition of the TPO when the likelihood of the project progressing is uncertain. Early discussion of values is a fundamental prerequisite, and this means bringing the key actors together to explore possibilities and discuss preferences, an approach central to the partnering philosophy and lean production ethos. It follows that the person responsible for putting the team together and implementing managerial frameworks has a crucial role to play. Selecting the 'right' project manager is therefore a critical first step.

Participants must be able to sign up to some common project values and common ways of working and commit to a continual search for improvement. This will be achieved only if activities are resourced in such a way that people feel valued and are motivated to contribute to improvements. Known as total quality management (TQM), the number one priority is customer (client) satisfaction.

Concurrent with achieving quality is the ability to see the process as an integral whole (not just as a series of individual activities or steps) and to recognise the link between these activities (processes) and the end result (the product). This is also known as process-orientated thinking. Following a TQM philosophy none of us should accept that the way we realise buildings is correct. We should be asking, 'Can it be done differently?' This means change (both incremental and radical), which is realised by empowered workers and effective leadership based on a clear purpose; it also implies measurement of performance and continual learning (see Chapter 9).

In some contexts it may be that the way in which the project is to be managed, the process, is already determined and it is a case of selecting the participant most suited to working with that specific process. Conversely, it may be possible to put the key contributors together before a decision is made on the process most suited to the participants and the project context. Either way, the synergy between the managerial processes to be used and the people working within the framework is crucial in ensuring an efficient interaction. Assembly of the temporary project organisation is about establishing a culture for effective communication and effective decision making, which is usually realised by assembling a mix of skills and experience.

Procurement choices

Procurement of architectural, engineering and construction services is paramount to the successful delivery of the client's goals and values. It is the procurement route that determines formal relationships, the nature and extent of participation, responsibilities, risk allocation and the potential value to be derived from the project. The client is faced with not only a variety of formal contracts from which to choose but also a wide variety of professionals from which to seek advice, all competing for the client's attention and apparently offering the best service. In some respects clients are faced with a chicken and egg problem. Which should come first, a decision on the procurement route or a decision on the participants that can advise them?

The initial choice of professionals will strongly influence the outcome of the project because of the social interactions that are set up at a very early stage. The lean philosophy of getting everything right at the start, before procurement routes and legal contracts are decided, should be noted here. The philosophy is about getting the right people, the right communication networks, the right supply chains, the right culture, and the most appropriate managerial framework(s) agreed very early. This helps to establish interfaces and transaction costs associated with the interfaces.

Choice of procurement route

Clients need to consider a number of factors before a decision on procurement is made, ranging from timing and flexibility to make changes, through to risk management, cost certainty and liability. The type of procurement route used will influence the manner in which the design and construction phases are organised, and hence how individuals interact and communicate through formal (and informal) communication channels. The type of system used will also dictate the type of contact used (see below) and hence the formal responsibilities of the organisations involved; this includes the organisations' individual level of control over the process, or parts of the process, and their exposure to risk. In some respects the choice of procurement route is about control and power over the project, information flow, communication routes, decision making and finance. Consideration of procurement routes should be done after, or concurrently with, the establishment of the project context. This will help to limit the choice of contract to be used. There are four fundamental approaches to procurement from a management perspective:

- Client-led relationships are common on very small projects (e.g. house extensions and some self-build projects). The client appoints consultants and contractors to carry out specific works packages through separate contracts (and on some self-build projects will carry out some of the works packages themselves). Communications are channelled through the client and interaction between professionals and builders tends to be rather minimal.
- Design-led relationships are known commonly as 'traditional' systems of procurement. The client appoints an architectural or engineering practice to design and oversee the construction of the building. Contractors are usually selected through competitive tendering, with the lowest tender being selected, although it is possible to enter into negotiation with pre-selected contractors and project partnerships. In this relationship, the architect or engineer is responsible for putting the TPO together and managing the project. In this type of relationship communications are channelled through the architect or engineer acting in the role of contract administrator.
- Construction-led relationships typically take the form of design and build (design and construct) contracts. Originally the contractor carried out the design work in-house and managed the construction of the building using directly employed labour. With the vast majority of main contractors now sub-contracting all of the work packages it is questionable whether this is a construction-led or a management-led relationship. In this relationship architects are dependent on contractors for their business and may have little or no contact with the building sponsor. Communications are channelled through the main contractor.
- Management-led relationships include management contracting, design and manage, and integrated (total) solutions. The majority of these aim to offer the client a 'one-stop-shop' service, which usually comprises a very large, multidisciplinary organisation or consortia that can provide expertise across all AEC project phases, from pre-project activities, design and construction, to post-construction and asset management. In this relationship a management team oversees the integration of all works packages. With increased use of prefabrication and off-site

production, and a reduction in the number of activities and trades to manage on the site, this form of procurement may be attractive to a growing number of architects and engineers and building sponsors keen to improve design quality. Communications are channelled through the project manager.

Choice of contract

Formal written contracts and agreements serve two purposes. First, they specify the rights, duties and obligations of each party to the contract or letter of agreement. This should be precisely and explicitly written, as poorly expressed contracts and letters of agreement are open to abuse. Second, contracts and letters of agreement serve as a reference document in case of dispute.

Construction contracts are usually categorised by their payment mechanisms, be it lump sum, percentage, unit price, etc. Whatever the contract chosen it will influence the allocation of risk and responsibilities between the project participants, setting the scene for the relationships within the TPO.

Attitude and trust

Contracts are concerned with formal communication routes, ideal situations that few participants follow to the letter (Emmitt and Gorse, 2007). Informal communication routes are required to make things run smoothly and are regularly used by project participants. Indeed, many actors may be quite ignorant of contractual arrangements until something goes wrong; they simply have a job to do. Although new contractual arrangements have been developed (and no doubt will continue to be developed), the construction sector still operates with different organisations (or different departments) dealing with specific tasks, hence actors have to communicate over organisational boundaries and communication barriers may exist at interfaces. The attitudes of the building sponsor and the team assembled to deliver the project will influence the choice of procurement route. The converse is also true. The procurement route and the contract will influence the attitudes of those participating. For example, it would not be unreasonable to expect a different attitude from the same organisations working with a project partnering contract compared with contracts based on competitive tendering. Similarly, concurrent working and fast-track construction will require a different attitude (and different skills set) from projects organised around sequential completion of tasks. Attitudes towards adversarial or relational approaches and the willingness of the client to become involved in the project will affect the behaviour of the actors and the ways in which they communicate.

Selection of the most appropriate participants

At the risk of introducing a touch of paranoia, managers would be well advised to think about the connections between groups and individuals in temporal project arrangements, and consider the probability of things going wrong. Alternatively known as 'Sod's Law', the principle that 'if something can go wrong, it will' has been around for a long time. It was not until 1949, however, that Captain Edward A. Murphy's name became associated with this truth (Matthews, 1997). He was the designer of a harness fitted with electrodes that was used to monitor the effects of rapid deceleration on aircraft pilots. During one of the tests on volunteers the harness failed to record any data. When Murphy investigated he found that every one of the electrodes had been wired incorrectly. This prompted the statement: 'If there are two or more ways of doing something, and one of them can lead to catastrophe, then someone will do it.'

Given the large number of professionals involved in AEC projects, and the often complex interconnections between them, it would be sensible to ask some simple questions relating to the assembly of the TPO very early in the project. The aim is to try and design the interfaces between organisations to keep them as simple and transparent as possible; and then not overcomplicate the connections with unnecessary bureaucracy and onerous managerial controls. Some of the questions that could be addressed include:

- Which professionals and trades are required to deliver the project?
- If we use a different type of construction technique, are more or fewer specialists required?
- When and how do the individuals connect with one another?
- Could those connections be achieved in another (better) way?
- Could the number of connections or interfaces be reduced or simplified?
- What could go wrong with the connections?

Confronting such questions as early as practically possible can help to ensure a good fit between the members of the TPO. Similarly, it can help to identify potential problems with some of the interfaces, which can then be mitigated and accommodated in the planning and scheduling of the project.

Applying the lean thinking philosophy to the assembly of the TPO can prove to be a worthwhile exercise and may result in questioning the role of some of the contributors. Is, for example, a cost consultant necessary on small projects when the function can be performed by computer software packages? Is a designer or structural engineer required when working with off-site, predesigned and prefabricated components and structures? There may not necessarily be a quick answer, but it is useful to ask the question and look at the value that the consultants and trades add to the project.

Selection criteria

Much has been written on the selection of staff to work in an organisation and organisational teams. Organisations have direct control over the type of people who make up the office culture and contribute to its profitability. New staff can be selected not only on their technical ability, but also on their ability to fit into the prevailing organisational culture of the office, i.e. their ability to interact and communicate with their new colleagues. In a project environment the composition of the TPO is dealt with by the project manager or a client's representative, resulting in a collection of organisations. Individuals are appointed to the project by an organisational line manager to represent the interests of each organisation. Rarely does the project manager have any direct influence on which individuals are allocated to the project by their employer. The result is that project relationships are indirectly imposed on (sometimes reluctant) individuals.

There is an old saying that you can choose your friends but not your family, and there is certainly some truth in this in AEC projects. When working on a project it is quite likely that we find ourselves interacting with individuals from other organisations whom we do not like, find difficult to trust and/or struggle to communicate with on a regular basis. Value management techniques and partnering initiatives seek to address this through careful selection of actors and team-building exercises. Working with competitive tendering may carry a higher risk of incompatibility within the project team, although this can be mitigated through pre-selection procedures.

With a shift from procedures to people has come a greater emphasis on the competences of the actors involved as well as their emotional (EQ) and social intelligence (SQ). Pre-selection or pre-qualification of project managers, architects, engineers and other key actors has become more widespread. In some cases project managers are asked to undergo psychological tests in an attempt to determine their suitability for major projects. In many cases the selection will be based on past experience, which is explored through a series of interviews to see if the individual is compatible with the client. Similarly, clients are paying greater attention to the people who will work on the project and it is relatively common for clients to ask to see the CVs of the main participants. They will be looking for a balance of qualifications, experience and evidence of continual professional development. Some clients will also ask to meet with the individuals who are likely to work on the project to look for social compatibility. Evaluating potential project participants is not just about their skills set, it is also about compatibility with other participants and the potential to work together, i.e. it is about matching competences and personality to specific temporary roles and tasks.

Depending on the project context and stage, selection criteria could include the following:

- *Attitude.* The attitudes of organisations and individuals to others (and the project in general) must be addressed to ensure compatibility with the project goals. Levels of trust and distrust are influenced by attitude; so too is risk.

- *Availability*. Are the individuals available for the full time frame of the project, or are they likely to contribute for a short period before moving onto other projects?
- *Communication skills*. Are individuals capable of communicating within and between disciplinary groupings in an effective manner?
- *Compatibility*. If organisational and individual values are not compatible it is likely that communications will not be as effective as they should be and the risk of natural conflict might be increased. Care is needed to ensure that the members of the temporary project organisation do not agree on every issue and have the confidence to challenge their fellow participants. Groupthink must be avoided.
- *Cost*. The cost of staff is an important consideration, and a mixture of experienced and less experienced contributors can help to keep the costs to an acceptable level.
- *Experience*. The individual's experience and the relevance of that experience to the project will influence the type and extent of interaction. In an ideal world a mix of experienced and less experienced participants would provide a well-balanced TPO. Project teams comprising a large number of inexperienced participants should be avoided because the lack of experience will result in the project team being ineffective. Similarly, teams comprising a large number of very experienced staff should be avoided, as they are inefficient, expensive and wasteful of resources.
- *Maturity and emotional stability*. The ability to cope with stress, unexpected events and pressure during the project and to be emotionally stable and consistent in dealing with others is a sign of maturity. This is particularly relevant to those in managerial positions within the TPO.
- *Motivation*. Motivation levels can influence interaction within the TPO, although this is difficult to gauge from an interview or CV.
- *Personality*. It is important that individuals are able to work together, thus individual personality clashes should be avoided. But this can be explored only through interaction, ideally team building workshops, prior to the commencement of the work. If it is not possible to arrange a workshop then selection is limited to the individual's reputation or experience of that individual from previous projects.
- *Qualifications*. Educational background and qualifications give an indication of an individual's academic ability. Evidence of lifelong learning will help to demonstrate an individual's passion for professional and personal development. Qualifications can help to establish whether or not professionals and tradespeople have the right qualifications for a specific context.
- *Skills*. Skills are demonstrated by educational and training qualifications and performance on previous projects. Skills relate to technical ability (as expected of the discipline) and social skills, such as the ability to

communicate across various levels within a project and the ability to work with other disciplines. Skills are usually demonstrated on previous projects, thus individuals tend to be evaluated based on their most recent project(s).

- *Values.* The discussion of values is also related to the roles undertaken by individuals and their individual and collective responsibilities. Early discussion and agreement of roles and responsibilities is necessary to avoid, or at least mitigate, problems later in the process. Consultants should be invited to discuss their values with the client and project manager. Appointment of the consultant should be done after some degree of compatibility has been established.

Selection of managers

The experience and personal characteristics of the managers will have a significant impact on the development of the project culture. Project managers (overseeing the entire project), design managers (responsible for design quality) and construction managers (responsible for the quality of the building works) occupy highly influential roles, and their allocation to a specific project should not be undertaken on a whim. Emphasis should be on appointing the most appropriate project manager for the project, and, as highlighted in the previous chapter, this first requires a thorough understanding of the project context. For example, a fast-track commercial project will need project managers with very different experiences and competences from those managing sensitive refurbishment work.

Selection of key individuals creates interfaces, boundaries between the individuals and the organisations that employ them. Individuals are usually assigned to projects by their employers. Sometimes experienced clients will ask to have specified individuals working on their projects because they have had a good experience in the past. Selection of key individuals to work on a project could entail one or more of the following techniques:

- *Shortlist.* This is drawn up based on education, qualifications and experience (and sometimes capacity to do the work) as listed above.
- *Interviews.* Interaction in formal and informal interviews helps to establish individual's social skills.
- *Psychometric testing.* Aptitude tests are used to measure intellectual capacity for thinking and reasoning. Tests are designed for a specific role and undertaken in examination conditions. They are usually used for senior managerial appointments at an organisational level, not a project level.
- *Workshops.* These are used to explore the values held by the participants with the aim of assembling a team of individuals who share similar values.

Organisational versus project resourcing

In an ideal world the most suitable people within an organisation would be selected to work on the project. However, it is common for the most suitable people to be already committed to another project (or projects) and so they may not be available. Thus less well-suited individuals may get allocated to the project, simply because they have spare capacity. The resourcing of projects can result in tension between the demands of the office and the demands of a project and/or client. Forward planning is crucial for the efficiency of the organisation as it collectively seeks to appoint the people with the best fit to new projects. So, it may not always be the 'best person for the job', but rather those least busy and hence most available to contribute. Tension between the demands of different projects within the organisational project portfolio must be managed with sensitivity so that the organisation makes the best use of its employees, the temporary project organisation benefits from the most suitable people, and individuals benefit from being given challenging and interesting projects to work on.

Outsourced relationships

In many project relationships one or more of the organisations will outsource work packages 'offshore' to help save money and time. This raises issues about how these outsourced relationships are managed and how (or if) they affect the dynamics of the TPO. Although organisations might argue that the benefits and challenges of outsourcing work packages offshore is a matter for individual organisations, not the TPO, it should be recognised that the work (information packages) produced will be used by the TPO, thus it needs to be considered in the wider project context. Work by Tombesi *et al.* (2007) has helped to highlight some of the technical challenges relating to the information produced and also to emphasise cultural and language differences in outsourced relationships, and hence the need for sensitive management. Much like project relationships, successful distant relationships rely on an assessment of each partner's contribution based on the specific context of the collaboration.

Building effective relationships

Successful managers understand how participants interact during AEC projects and how they communicate within the context of the TPO. Time spent on creating and maintaining contacts can be an effective strategy. Communication networks, the system architecture of the project, need to be designed with the same care as that expended on buildings. This involves implementing appropriate ICTs and trying to assemble the most suitable people for the project. These tasks should be undertaken as soon as practicably possible after a decision to engage in a project has been taken. In some situations it is possible to consider the system architecture before briefing starts, although it is more usual to do this concurrent with the early client briefing stages. It is also common to leave this until later in the process, which means that many of the connections are already made and the communication networks tend to develop organically, which may make the management of the system more challenging than it needs to be.

Start-up meetings or (facilitated) workshops may be used to bring together representatives of the main stakeholders (see also Chapter 4). These early meetings should include the client and/or the client's representative, architects, engineers, project managers and, if known at the time, the main contractor and/or specialist contractors. Representatives of user groups may also be present for some building types such as social housing. These meetings should be used to explore the values of the stakeholders. Various approaches are taken to start-up meetings and workshops. It is common practice for the project manager to take responsibility for arranging and directing the meetings and workshops. Another approach is for an independent person (someone with no contractual responsibility for the project) to act as a facilitator. The facilitator's primary aim is to encourage open communication and the development of working relationships based on shared values and mutual trust. This takes time and a number of workshops and activities may be needed to help build a team spirit. Subsequent workshops should focus on improvement of team interaction and further development of relationships based on trust.

The intention is to build the system architecture for the project, thus allowing actors to engage in open and effective communication during the life of the project. In addition to trying to engineer the project culture and hence make the events that follow run smoothly the outcome of many team building events and workshops is the signing of a partnering agreement (or similar document). Such agreements commit the project stakeholders to a working relationship based on open communication, knowledge sharing and trust.

Project cultures

Much has been written on the subject of organisational culture and these ideas and theories are often transferred to a project environment. Organisational cultures are dynamic and constantly changing as organisations evolve to suit their business environment. However, they are relatively stable environments with stable membership compared with a multidisciplinary project environment. The project culture is a dynamic interaction of many organisational cultures, which are constantly shifting in response to their business environment and their required input to the project. Therefore, to talk about a 'project culture' may be a little misleading, as the term implies something that is relatively constant throughout the life of the project, which it is not.

A contracting organisation may have a strong corporate image that has been developed over a long time period and reinforced through individual projects and marketing initiatives. By comparison, a TPO will not have a corporate image at the outset; instead the TPO will borrow organisational cultures from the main contributing organisations, effectively forming a swift culture. This swift culture will evolve as the organisations start interacting, producing a positive or negative culture in response to the quality of the interactions.

'International' projects

Working internationally may help to reduce the exposure of an organisation to local downturns in the economy, but it can also present additional challenges in terms of working across unfamiliar cultures and languages. Although collaborative communication technologies can mitigate the need for extensive travelling, it is still necessary to understand the context in which the project is being delivered. Sometimes, even slightly different interpretations of words or roles can cause difficulties, especially if they are not identified at an early stage in the project. Research into the performance of international project organisations has helped to identify some of the difficulties faced. These range from language and work culture through to the use of unfamiliar building components and technologies, all of which can affect the performance of the project (Grilo *et al.*, 2007).

Establishing a project culture

Early establishment of values through discussion can help to establish compatibility, start to develop trust and establish interpersonal communications. This should relate to the project objectives, which should also be reviewed at the team assembly stage (Figure 7.1) It is through discussion that a project philosophy can be established and communicated to new members of the TPO as they enter the project. Collectively, by addressing values, objectives and philosophy, a project culture will start to emerge (Figure 7.1):

- *TPO values*. These comprise both individual values (which need to teased out, but are likely to remain hidden) and project values (which need to be articulated and agreed collectively).
- *TPO objectives*. These comprise individual, organisational and project objectives.
- *TPO philosophy*. Each project will have its own philosophy, often driven by the client and key advisors. This can influence the attitude of the TPOs members.

Early project meetings tend to involve senior members of the participating organisations, but it is not these individuals who will do the work. Indeed, in some projects they will be rarely seen again unless there is a major problem.

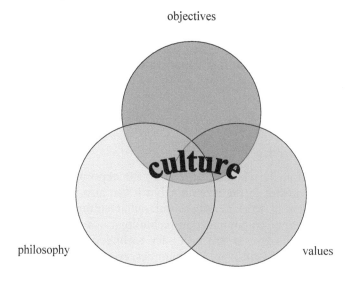

objectives

philosophy

values

Figure 7.1 An emergent project culture.

Instead the work is delegated to others of varying degrees of seniority within the organisation. The implication of this is that the team-building exercises need to be carried out on several occasions to ensure that the most active participants are present. On projects with a long duration of several years and with distinct phases it may be necessary to engage in team-building exercises on a number of occasions, strategically placed within the overall project schedule. This will add to the costs associated with the management of the project, but, if done well, will help to encourage effective communications and may go some way to reducing natural conflict.

Practical challenges

The most common challenge is associated with starting the project before the main project participants have been assembled, perhaps in the hope that the TPO will build itself into a dynamic and efficient entity; it will not. Guidance and a considerable amount of effort are required to ensure that the project starts from a sound foundation. Although this does not guarantee that the project will run smoothly, investing time in designing the interfaces and hence the formal communication routes should be beneficial in the long run, saving time and money as the project develops. This can often mean discussing the programme with the client and delaying the start of work packages until the fundamental issues have been resolved. This can involve some difficult negotiations with clients, and the ability to educate clients about the potential problems of starting too early is an essential attribute.

This is illustrated in the following story taken from a very large project that was procured under a partnering agreement.

Construction started before the design had been adequately resolved. All of the main consultants and the principal contractor were aware of this. The problem had been highlighted by the project managers to the client, asking for a two-month delay to the start of construction, but the client chose to ignore the advice and proceed. Because the conceptual design had not been resolved into sufficient detailed information the programme soon started to experience pressure: the contractor could not start certain works packages. Partly because of the pressure brought to bear by the principal contractor on the design consultants, combined with the designers' inexperience of very large projects, negative conflict soon developed, which merely added to the problem. Delays were starting to affect the cost (which was rising quickly) and the morale of the individuals working on the project (which was falling quickly). The client, conveniently forgetting the advice not to start, insisted that the programme had to be met, and so the project managers introduced a number of 'new' management ideas and additional project managers in an attempt to resolve the problem. The management interventions, although well intended, failed because the participants had already decided that the project was a disaster. Interfaces had broken down, communications were problematic and, despite the existence of a partnering agreement, trust and goodwill had evaporated.

With the benefit of hindsight the project managers involved with the project accept that it would have been better, and more efficient, to stop the project, remove the organisations that were not performing, resolve the outstanding design issues, appoint appropriate consultants and start again. They claim that requests were made to the client to stop and resolve the issue, but that this was overruled. It was estimated that to stop the project and resolve the problems would have taken around three to four months; in the event the project overran the original programme by 15 months. The individuals involved in this particular project were rather bitter about their experiences, the client lost a considerable amount of money and the organisations involved also claim to have lost money. Because costs had mounted a number of cost reduction techniques were put in place and the finished building was built to a lower standard of finish than that specified. Altogether a rather unsatisfactory state of affairs.

End of chapter exercises

- You are approached by a client and asked to work on a derelict building. There are two options that could be pursued: (1) demolish and rebuild or (2) restore and reuse the building. The client has asked you to propose a TPO for each situation and explain your reasoning for the selection. Who is required, and what are the main differences between the composition of the two TPOs?

- You are assigned to a project and find that the main contact in one of the organisations is very difficult to communicate with and furthermore there appears to be an issue with compatibility between the two of you. What could you do about the situation?

Further reading

Readers will need to consult literature on procurement and contracts that relates to their location and/or project context. Generic guidance on assembling TPOs can be found in a wide range of project management books.

8   Development

Of equal importance to the assembly of a temporary project team is the ability of managers to keep the TPO functioning effectively during the various stages of the project. Relationships evolve, conflict manifests and levels of trust are constantly being tested and redefined. Development of the TPO is challenging because of the large number of disciplines contributing over a long time frame. As the project develops some participants leave the project, their task(s) complete, and others join, and in doing so affect existing patterns of interaction. Some participants will leave temporarily, to return later in the project life cycle as their contribution is once again necessary. Gray and Hughes (2001) have attempted to describe this phenomenon, which they call the wheel of dominance, with actors being active at different times. Although the wheel is a useful metaphor it is potentially misleading because rarely do the participants interact with the regularity of a revolving wheel. Instead it may be useful to see the varying levels of participation (and hence dominance) as planned insertions in accordance with the project plan, and unplanned insertions in response to unexpected events.

AEC projects have an ephemeral character, with the temporariness of the arrangement prone to breaking apart. Experienced managers will readily attest to the rule of thumb that it takes about one per cent of their time to assemble a project team and the remaining ninety-nine per cent of their time to keep it functioning effectively, i.e. to prevent entropy. The project manager and key actors must put considerable effort into maintaining informal communication and effective collaboration, a task of equal importance to the achievement of individual work packages and tasks. An appreciation of group norms and how changes in membership can affect performance can be useful in helping to nurture the project team and guide it to a successful conclusion. Similarly, the ability to reassess the context of the project on a regular basis, built into formal review events, will help to reveal the changes happening as part of the project's evolution.

Fundamentals

A relatively free and open atmosphere characterises interactions during the early stages of a project. Everyone is optimistic and positive and the amount of information in the system is relatively low and largely unstructured. As the project develops, the level of information increases and relationships mature, and informal and formal structures emerge, which creates implicit rules of engagement. Goals and targets are set, reappraised and adjusted as work packages are completed quicker or slower than scheduled. As milestones and deadlines become more urgent the pressure to develop, exchange, coordinate, complete and approve information mounts. This can, and often does, place considerable pressure on individuals to perform and may test often fragile relationships, sometimes resulting in conflict. The pressure resulting from limited time, a large amount of information and increased workload can affect the ability of team members to work with each other, leading to changes in communication patterns and levels of trust. This may either impede or help to improve group performance.

Levels of enthusiasm also vary. A well-known characteristic of projects is the enthusiasm and energy that greets new projects and the tendency of people to drift away towards the end as other projects within their organisation become more important (and perhaps more profitable), and hence more demanding of their time. Enthusiasm levels can also wane on long projects as individuals become jaded and tired. This is common to all projects and stages of projects, regardless of sector. To counteract this phenomenon contracts usually have a retention sum to 'encourage' contractors to finish the work to a satisfactory standard before moving on to another project; however, most managers experience difficulties in getting people to finish their work (especially if that work is relatively minor compared with the overall work package). Here the motivational skills of the managers are important, as is the self-motivation of the contributors.

Relationships form, evolve and disband throughout the life cycle of the project. This is an important point to make, because newly formed groups exhibit different characteristics from established groups. In a newly formed group, members must simultaneously find their place within that group and socialise with one another. Each AEC project embraces a new grouping of actors who must attempt to develop social relationships that support the project. Research on group interaction and socialisation processes found that in new groups the exchanges present a two-way process of social influence, changing other's ideas and opinions through communication. It is claimed that the norms that govern group members come into force as each of the actor's objectives become apparent. The group norms regulate group behaviour. As groups develop their members get to recognise the importance of individual professional roles. The group regulatory forces are then used to control interaction of individual members. In construction, as the tasks change from feasibility to design and then construction, the power, importance and involvement of the professionals will change. The challenge for project managers is to try and lead this dynamic process. As the project develops and the specific demands of the situation change, the actors with the most relevant skills become more influential and powerful. Those with relevant skills emerge or are nominated by other members of the group to be the most dominant for a particular period of time, during which they are central to the information processing and decision making because of their expertise.

Group development

The individual development of the many groups within the TPO will have an influence on their ability to interact with other groups and hence the overall performance of the TPO. Similarly, the strength of the professional relationship between designers, managers and trades, and their ability to resolve problems, will also be a function of their interaction behaviour. Achievement of a group's goals depends on concerted action and so group members must reach some degree of consensus on acceptable task and socio-emotional behaviour before they can act together. Interaction has been described as task-specific and social. The social element of interaction is developed through emotional exchanges that are used to express a level of commitment to the task and other members. To accomplish group tasks relationships need to be developed and maintained. The level of interaction associated with maintaining, building, threatening and breaking up relationships will be a function of socio-emotional interaction and will be subject to group norms.

A group's behaviour develops and changes over the period of interaction. As task groups attempt to resolve the challenges allocated to them, they undergo changes in terms of their attitude and behaviour towards each other (Hoffman and Arsenian, 1965). Groups go through a process of learning, which can result in changes to structure as they move through a range of social, emotional and developmental stages (Bales, 1950). Two variables affect group development: the length of time that a group has existed (Lieberman *et al.*, 1969) and the number of occasions that the group has previously met (Hoffman and Arsenian, 1965). Both of these variables are important considerations in a AEC project context. When people have taken part in a series of meetings on related subjects and different people were present in each of the previous meetings (a common occurrence, for example, in construction progress meetings), group participation is the same as if the group had met for the first time. This phenomenon is due to the group's socio-emotional development. Individuals are not aware of the group's social and emotional norms, nor does the group know how the newcomers will react to the group norms. Thus, a socio-emotional framework develops and re-establishes itself when new people enter the group.

Early group meetings necessitate the formation of a socio-emotional structure and a participation strategy that will be used to access information, develop group knowledge and make decisions (Shadish, 1981). When actors have experience of the subject being discussed, but have not experienced interaction in that group setting, they are more likely to use task-based interaction than socio-emotional interaction. Socio-emotional communication emerges later once the newcomers are familiar with the group members and the group norms. Agreeing and disagreeing are socio-emotional communication acts. Equally, emotion is used to show support and concern. In groups it is the socio-emotional interaction that regulates interaction and provides the framework against which decisions are made.

In a project environment not all teams are 'new' and group norms and behaviour may be influenced by previous relationships. Although each AEC project is unique, some groups will maintain relationships at the termination of a project (or project phase) with the aim of working together in the future. Clients, developers and consultants usually have a number of preferred suppliers and personal contacts (individuals within organisations) that they will use if given the freedom to do so. Alternatively parties may contact those they have previously worked with for advice or help without entering into a contractual arrangement. Working with known actors through informal strategic alliances or formal strategic partnerships may save time and improve knowledge transfer because relationships have developed, a level of trust exists, some formalities have been dispensed with and the skills are known (and to some degree have already been tested). In these situations emphasis will be on group maintenance and development activities.

Work-oriented groups need to maintain a balance between task and social demands. As groups address problems emotions start to develop and, as a result of disagreement, tension is built up between members as they focus on the problem rather than relationships. Conflict, even when constructive, leads to tension that can damage the cohesiveness of the group and threaten group maintenance; yet too much attention to cohesion tends to stifle constructive conflict, can threaten the group's ability to solve problems and may lead to groupthink. For a group to be effective, task issues must be discussed, and it is to be expected that conflict will emerge in the process. Functional conflict can help to avoid groupthink and improve the decision-making process, but conflict may also damage relationships between group members if not managed diplomatically. Relationships must be managed so that they are sufficiently sustained to bring the discussion to a collectively agreeable conclusion while maintaining the harmony of the group.

Development of group norms

Although the behaviour and characteristics of groups change and develop over time, it is well known that groups develop (and are subject to) behavioural norms. Newly formed groups will develop relatively stable patterns of interaction as they mature (Heinicke and Bales, 1953; Keyton, 1999), demonstrated through recurrent patterns of thinking and behaviour. Anderson *et al.* (1999) make a distinction between rules and norms, noting that members come to accept norms as their way of being a group and doing group work, whereas rules are agreements about how to behave appropriately. The norms of group behaviour may be specifically associated with the reason why or purpose for which the group formed, the task, or they may be attributable to the group make-up.

Group norms may change as members adapt to changes in (project or organisational) context. In almost every situation there are a number of specified roles, environmental clues or repertoire of acts that provide information about how the individual is suppose to interact (Jackson, 1965), and these vary from one situation to another (Furnham, 1986). Expectations of the way that members are supposed to act are translated into implicit rules that are adopted by the group to regulate its members' behaviour (Feldman, 1981; Jackson, 1965). Norms and rules provide powerful controls over the group, although it is those that are implicit that have the greatest direct effect on relational behaviour (Keyton, 1999).

Group norms can be so influential that some members will express a judgement differing from the one they hold privately (Hare, 1976). Feldman (1981) has identified four ways that group norms are developed:

- Norms can develop from behaviour and statements made by group leaders.
- Critical events in the group's history can establish a precedent, for example in how it responds to a deadline or unexpected event.
- Norms may simply develop from repetitive behaviour patterns, such as the way in which meetings are chaired.
- Members can import group norms from previous group experiences.

Newcomers need to observe the overt behaviour and practices of group members so that they can understand the group culture and participate in it (Trujillo, 1986). When new groups form they establish beliefs, values, norms, roles and assumptions as a result of communicative behaviours that uncover similarities and differences (Feldman, 1981; Anderson *et al.*, 1999). An individual's actions and behaviours are also influenced by his or her motives for membership, positions and role (Zahrly and Tosi, 1989).

Group goals and productivity

Groups are more than the sum of their component parts (Hartley, 1997). Groups that have been found to be more productive than others are said to have a structure that is suited to their function (Hare, 1976). The high level of productivity is achieved not only because they have procedures for solving problems, but also because the group is stable and less time is devoted to status struggles (Heinicke and Bales, 1953; Hare, 1976). Similarly, the members are aware of each other's skills, attributes, knowledge and roles and only a relatively small amount of discussion is required to organise tasks. Over time the knowledge of each member's skills and attributes should make the group more effective. Roles and responsibilities can be assumed and the most appropriate person can be quickly allowed to undertake the task, without the need for lengthy discussions to determine who has the necessary skills or knowledge.

Bales (1970) suggested that role ambiguity may lead to apathy towards overall goals, disrupting the early stages of group development and interfering with the implementation of a compatible social structure for the group. Establishing the aim of the group and the role of the individual and developing a group structure that aligns individual skills towards the goal is important. Incompatibilities and ambiguity will cause problems and so conflict within the group may be a natural way of (re)aligning group behaviour or conversely disrupting the group as it develops and changes and new information is considered by group members.

Effectiveness of groups and the degree of cooperation between their members will depend on the nature of the communication strategies employed (Ackoff, 1966; Hollingshead, 1998) and the training provided (Gutzmer and Hill, 1973). According to Hare (1976) one of the clearest findings in the literature on small group behaviour is that the productivity of the group, no matter what the task, will improve if training is provided. However, Morgan and Bowers (1995) have argued that training must be suited to the context otherwise training programmes may have a detrimental effect on group performance.

Inconsistent membership

Given the complexity of AEC projects and the tensions between the demands of organisations and project, it is not surprising that the TPO rarely has a consistent membership. Uncertainty about who will be representing a particular organisation at a particular juncture in the life of the project will always be present because it is not possible to 'fix' individuals to particular projects. Line managers may need to move personnel from one project to another to meet the demands of their organisation, or they may need to fill a vacancy created by, for example, a prolonged period of sick leave. As individuals leave and new members enter the TPO it is essential that the project philosophy is communicated to the newcomers. This can be done as part of the induction process, at formal handover meetings and as part of the health and safety induction, for example when new trades enter the construction site.

Whatever the reason behind changes in personnel the result is inconsistent membership within the TPO, resulting in the need to re-establish communications and relationships with the new individuals, as illustrated in the following case studies. These have been chosen because they help to highlight the problems of changes to key team members during projects. These stories are not particularly unusual and most practitioners can readily recall similar events. What is difficult to know is how frequently changes are made to TPOs and what impact this has on project outcomes.

Case study 1

An architectural practice was appointed directly by the chief executive of a large organisation to work on a relatively small, yet complex, project that involved alterations to the organisation's head office. Given the importance of the design the chief executive and other top managers were closely involved in the development and approval of the design brief and the conceptual design. Following approval of the conceptual design the chief executive then handed the project over to a middle manger, who became the main (second) client contact. Up until this time relationships had been harmonious and the project was progressing well. The middle manager was highly personable and committed to his job, the only problem being that he wanted to stamp his mark on the project and was continually requesting changes to the design, both before and during construction. This caused the architectural practice a small amount of rework, although the relationship was positive and resulted in a high quality design. Shortly before the project was completed the client representative announced that he had been promoted within the organisation and that he would be replaced. His replacement, the third client contact, was not as approachable and was not happy with the design. A large number of changes were requested, despite the fact that the design had been approved and most of the work had been completed. This resulted in a dispute between the client contact, who had not had any previous input into the project, and the architectural practice, which marred an otherwise successful project from the contributor's viewpoint. Eventually, after some difficult discussions some of the changes were made post project completion, at the client's expense.

Case study 2

This example was taken from a relatively simple housing project comprising apartments and communal spaces. The contract had been awarded to a large contractor on the basis of a competitive tender. During the twelve-month contract period there were no fewer than four project managers assigned to the project by the main contractor. The first project manager stayed with the project for three months, during which time the project was progressing well (on target for time and cost, and no reported problems). Because this project manager was doing a good job his line manager moved him at short notice to a different site that was experiencing problems. The project continued without a project manager for three days, which resulted in some uncertainty amongst the trades working on the site. He was replaced by a second project manager who stayed for four weeks and two days. Project manager number two had very little experience of housing projects; he had just finished a large commercial project and was waiting for a new commercial project to start. When it did, he was replaced by a third project manager. During this period the project had 'drifted' and a number of small, but collectively significant problems had mounted. The third project manager's approach was to intimidate the sub-contractors to try and get the project back on programme. Not surprisingly, this approach had a negative effect. Two of the sub-contractors walked off site and refused to return until the project manager was replaced. Eventually, after three months, the project manager was replaced by the main contractor, although not until a lot of damage to working relationships had been inflicted. The fourth project manager stayed with the project for the remainder of the contract. His first job was to try and rebuild the damaged relationships.

It appeared that the main contractor was using this project as a holding project for their project managers. This was denied by the main contractor, but this strategy was 'quite common' according to the fourth project manager. Not surprisingly, the project was completed two months late and was significantly over budget. Although shuffling project managers appeared to suit the contractor, it was the tradespeople who suffered the consequences, and the client who paid for it. Interestingly, there was no mention of the change in project managers within the minutes of the progress meetings (merely a different name next to the project manager position). The contractor claimed a whole raft of excuses for running late, blaming the weather, the architects, the engineers and the sub-contractors for poor performance, when in fact the problem appeared to stem from inconsistent management by the main contractor.

Shuffling personnel also happens in architectural and engineering offices, with staff being moved from one project to another in an attempt to best allocate resources from the perspective of the organisation – which may not necessarily be in the best interests of the project or the client.

Conflict

Conflict first emerges when an individual feels that someone else has frustrated, or intends to frustrate, some concern of theirs (Hargie *et al.*, 1999). This perception of conflict can result from differences of opinion, simple misunderstandings, mistakes and/or fundamental differences in requirements. Thus conflict exists where there is an incompatibility of interests. One way of interpreting conflict would be as a breakdown in communications, an inability to explain or direct at the appropriate juncture, thus leading to frustration and in the worst cases escalation into a dispute.

Conflict is an inherent feature of work groups with an important role to play during problem-solving processes. With so many people involved in AEC projects it is highly likely that there will be some form of conflict during the project. Conflict has been found to develop in AEC projects as group members discover their team objectives and then attempt to enforce them on others. Conflict in projects is also associated with the control of scarce resources (finance, labour and materials), which often manifests when changes are introduced to the design, budget or programme, resulting in power struggles.

During a group's development a relatively defined structure of interaction evolves through the group's regulatory procedures. Through experience, group members learn to expect conflict in certain areas between certain members, and also expect to gain support from others. As the group's socio-emotional awareness develops, the members anticipate where conflict could occur and use supportive reinforcement interactions to overcome it. This allows participants to engage in task-related elements and control discussions with socio-emotional interludes. People tend to be more inclined to express their emotions when they are more familiar with each other, thus the longer groups have been together the more likely negative emotions are to be shown. Established groups may not necessarily experience less conflict than newer groups, but they tend to be better equipped for dealing with it.

Types of conflict

In very simple terms there are two types of conflict, natural and unnatural. The problem is that it is not always easy to identify the difference between the two:

- Natural conflict is the intended or actual consequence of an encounter, resulting in stronger participants benefiting from the clash. This is inevitable and thus some plan to deal with it can be made in advance.
- Unnatural conflict is when a participant enters into the encounter intending the destruction or disablement of the other, usually with the intention of making a financial or personal gain. This is quite a well-known strategy of less scrupulous contractors looking to increase their profit margins on a project and something that project managers need to be wary of. Unnatural conflict can occur in organisations and TPOs as individuals seek to benefit from undermining their colleagues.

The conflict life cycle

For conflict to develop there needs to be a trigger event. The trigger may be caused by incompatible behaviours, personal dislikes, scarce resources, unexpected changes, incomplete information or differences in expectations, i.e. a disagreement. Underlying characteristics are likely to be related to differences in issues, perspectives, understanding, beliefs, values, goals, opinions and needs. When conflict emerges the parties involved will need to engage in discussion and negotiation, which leads to tension as the disagreement is addressed. Successful conflict management is able to explore the differences and reach a settlement without damaging the relationships between the parties. If the conflict is not adequately resolved and the tensions remain then the consequences are undermined relationships and weakened business relationships. Therefore, the conflict life cycle starts with disagreement, moving to confrontation and escalation, before moving to de-escalation, resolution and aftermath (Emmitt and Gorse, 2003):

- *Disagreement.* As members of the TPO discuss task-related issues it is inevitable that at some point there will be a difference of opinion. Sometimes this is expressed to others, sometimes it is withheld because individuals are not confident or prepared to state their opinions. It is necessary that disagreements are voiced, although it is expected that some tension will arise from the discussion that follows.
- *Confrontation.* When the opinions being discussed are incompatible the confrontation stage begins. Although attempts will be made to voice beliefs and facts rationally and calmly, people often feel passionate about their cause and this can lead to the use of emotive language and body language, which can increase levels of tension and anxiety.
- *Escalation.* As tension builds, the conflict (not the issue that brought it about) becomes the focus of attention and individuals seek to protect their self-image rather than continue with the original topic of debate. If the situation is not managed, i.e. controlled, then the result will be an increase in misunderstanding, tension anxiety and mistrust.
- *De-escalation.* Conflict can make people feel uncomfortable, can be tiring and takes time that could be spent doing other tasks. Therefore it is in the interests of individuals to try and resolve the conflict quickly. By reiterating common goals and encouraging rational discussions it may be possible to control emotionally charged exchanges. Similarly, praising individuals for their constructive input and apologising can help to get discussions refocused on the disagreement.
- *Conflict resolution.* Resolution to the disagreement will occur when all involved parties agree to a proposed solution, i.e. there is a degree of compromise. Conflict may also be resolved if individuals withdraw from the situation and/or agree not to work together.
- *Conflict aftermath.* Short-term consequences will be reflected in decisions relating to task-based decisions. If the conflict has been resolved in a positive manner the basis for a stronger relationship is formed; however, if individuals feel aggrieved by the outcome then it may undermine relationships, often having long-term repercussions.

Conflict management

Numerous publications have claimed that conflict is an unwanted characteristic of the AEC sector that adds unnecessary costs to projects, yet research has shown that certain types of conflict may be beneficial to the development of projects (see Emmitt and Gorse, 2003). It is not conflict *per se* that is the problem; rather it is the indiscriminate (or poor) management of conflict that causes the difficulties and leads to lengthy disputes that are expensive to resolve. It follows that there is a need for all contributors to AEC projects to recognise the signs of conflict and manage conflict to the benefit of the project. The ultimate aim is, of course, to prevent the conflict escalating into a legal dispute.

Conflict needs to be managed so that it does not suppress information or become personal and dysfunctional and damage relationships. Most conflicts are managed by exploring alternative solutions, different perspectives and encouraging all participants to engage in discussions and, hopefully, reach agreement. Conflict may be beneficial or destructive to the performance of groups and teams:

- Benefits of conflict include increased understanding of issues and opinions, and greater cohesiveness and motivation. When group members disagree and explore why they disagree they expose key issues and points of misunderstanding. Groups that experience tension and conflict often feel closer and stronger after working through a crisis.
- Disadvantages include decreased group cohesion, weakening of relationships, ill feelings and destruction of the group. Conflict between people can be distasteful and personalised, having little relevance to the task or problem. Most people do not like to be criticised and all conflict has a negative socio-emotional impact that must be recognised.

Resolving disagreements

When conflict takes place it is important that the impact on individuals is tempered with some form of positive socio-emotional support. The support does not need to come from the person who initiated the criticism, but the group's relationships must be managed. Groups that do not work through conflict and repair relationships will fall apart. Some conflicts may turn into disputes and this involves considerable cost and inconvenience for all parties. The negative perceptions that develop from public dispute often serve to damage all parties.

A certain amount of conflict within any organisation is inevitable and the existence of communication problems will make the management of conflict difficult. The management of conflict within organisations needs to concern itself with the reduction and eradication of dysfunctional conflict and the manifestation and management of functional conflict, which will help discussions to remain creative and useful. When there is more than one organisation affected by conflict, such as in a project environment, each organisation will attempt to secure its own goals before addressing those of the temporary project organisation. A supportive group climate should be developed so that when conflicts emerge, as they inevitably will, the group is able to repair emotional damage and continue with the group task.

Conflict often emerges from perceived failure. When the failure level is high there is a greater chance that people perceive the task to be impossible or that the chance of failure is so high that it is not worth the effort. Very low levels of failure may be taken as a job achieved, but not done too well. Moderate failure, in which the task is considered tough but achievable, may present a challenge and a chance to prove to oneself and others that the task can be achieved. Thus, moderate levels of conflict (being related to perceived failure) may be productive. Negative feedback can be stressful and group members need to be aware of the development of socio-emotional tension. Debate and negative emotional exchanges may threaten relationships; however, negative emotion can be useful if positive socio-emotional interaction is also used to maintain relationships. Any socio-emotional tension that develops is removed by positive emotional acts (such as showing support, joking and praise) and negative emotional acts (such as disagreements, an expression of frustration and even aggression). If socio-emotional issues are not addressed in a timely fashion the increase in tension may inhibit the group's work. Groups must maintain their equilibrium, moving backwards and forwards between task- and socio-emotional-related issues. Negative and positive socio-emotional interaction is interlinked with the group's task-based interaction. The task-based interaction will provide information about possible ideas, actions and directions of the group. The positive and negative emotional signals will provides clues about how group members feel about the suggestions and encourage others to provide further information. Through suggestions and ideas, the subsequent testing of them and the reaction of the group (conveyed in rational and emotional responses), members are able to learn what acceptable behaviour for the group is. Too much attention to task interaction may limit the communication required to build and maintain relationships. If groups are to perform effectively positive reinforcement (agreeing, showing solidarity, being friendly and helping release tension) is needed to offset negative reactions (showing tension, being antagonistic, appearing to be unfriendly and disagreeing). The majority of project teams will need help, thus separate team-building activities need to be included in the project schedule.

Group members feel comfortable in a positive socio-emotional environment. Members prefer positive feedback, and interaction which suggests that the group is effective can help to increase morale, but too much may be counterproductive and result in groupthink. Socio-emotional interaction is important in the accomplishment of tasks. The combined effort of the group requires exchanges that develop and maintain the group and provides a social structure capable of decision making and task accomplishment.

Occupational stress and burnout

Individual workload will influence enthusiasm levels, the perceived level of occupational stress and the tendency for individuals to become exhausted. Within the TPO it is highly probable that participants will experience some difficulties with completing their work on time, may experience occupational stress and may become tired and uninterested in the project for a while. This may be a direct result of the demands of one project, although it may be that individuals are working concurrently on other projects that have conflicting demands. Managers within organisations need to give reasonable work allocations to their colleagues. Project managers also need to be alert to these challenges and be prepared to talk to individuals in an attempt to avoid serious problems.

A common characteristic of projects is that individuals tend to be highly motivated and enthusiastic at the start, but enthusiasm levels tend to decline over a period of many months and years as the project progresses. At the completion of projects, or the completion of major work packages, it is common for individuals' enthusiasm levels to be low, especially if they have already been allocated to other, more exciting projects. This may be linked to tiredness and burnout, which may in turn be linked to sustained periods of occupational stress.

Occupational stress

It is inevitable that we will experience some stress in our daily interactions with others.

Stress is a result of the way that we internalize and respond to external events. Whereas the level of stress perceived by an individual is related to the event, the way in which we view events and react to them will also contribute to our stress levels. Thus some people appear to be able to cope with more pressure than others in a similar situation.

Stress is linked to uncertainty and manifests itself as an uncomfortable mental state. For example, stress can come about if an individual is given a task to do that lies outside their area of expertise or is not given appropriate authority over decision making. The result is that individuals become less effective and may suffer poor health.

LeDoux's (1998) work on emotion and the brain provides a useful insight into how people process information when under stress. During stressful situations the adrenal gland secretes a steroid hormone into the bloodstream. The release of the hormone helps the body to mobilize energy resources so that we can deal with the stressful situation. During particularly prolonged or very stressful encounters the brain fails to regulate the chemical release, and excessive levels are released into the blood. The chemical overload causes the brain to work differently and we may become unable to remember things, experience difficulties in learning or struggle to make decisions. Prolonged or very stressful situations can result in permanent brain damage.

Burnout

Burnout is a term used to describe an individual who becomes tired and exhausted because of work demands, leading to cynicism about their work and inefficiency (Freudenberger, 1974; Maslach, 2003). Burnout is also likely to result in a decline in social interaction and the inability to complete work packages effectively and efficiently. Although there is no clear consensus about the link between occupational stress and burnout, it would appear that burnout is more likely to occur if the individual is not well matched to their work. In the majority of cases a short break or holiday will address the issue, but in more serious cases the individual may need longer to recover.

Project closure

Project closure is the final stage of a project during which all project objectives should be completed and the resultant artifact handed over to the client (customer). Within the AEC sector project closure is rarely absolute, with projects and phases often handed over with (minor) work still to be completed, usually listed in the defects (snagging) list. The effect of this is that relationships will continue for some time after the project has been handed over to the client. Thus the TPO will continue, albeit in a much depleted state, to help facilitate the completion of unfinished work and the correction of defective work.

Ending projects or completing major stages of projects can be challenging and is a stage when the pressure to complete work can, if not managed well, result in conflict and disputes. In situations in which the project has exceeded its resource base, for example time and/or cost, there will be additional demands placed on the participants and this can sometimes place too much pressure on individuals to perform. At this final stage in the life of a project it is not uncommon for relationships to deteriorate as the contributors try to establish who is responsible for the outstanding work and who is responsible for defective work – issues that are rarely straightforward given the complex interfaces between contributors and materials/components. Project managers will usually need to exercise considerable tact and diplomacy to encourage the participants to complete the project in a harmonious and positive manner. Successful completion of the project will be a cause for celebration. Unsuccessful completion will usually result in the application of resources to resolve disputes for many months after the completion of the project, a situation that adds no value to anyone.

One of the obvious effects of finishing a project is that the TPO ceases to exist, and in the majority of cases individuals find themselves working with new organisations and individuals on their next project. Thus the whole cycle starts again, but with a new context and new interfaces. It is essential that the project is evaluated, not only on the standard performance criteria of time, cost and quality, but also on the way in which the interfaces were managed and the quality of interaction within the TPO. Most project plans include a project closure report and a post-project review. The project closure report will address issues specific to the project as well as a mechanism for informing all stakeholders that the project is completed and hence closed. The post-project review enables the key stakeholders to analyse the outcome of the project to determine whether the project delivered the value expected by the client. This review mechanism also allows an analysis of the project delivery system.

In addition to reviewing the performance of the project and the resultant artifact all contributors should try to learn from the project, both the good and not-so-good experiences. This should help to mitigate or avoid similar challenges arising on future projects. A number of tools and techniques can be applied to assist with learning, which are discussed in the next chapter.

Practical challenges

The factors surrounding group development are complex and there are no easy recipes for ensuring that interdisciplinary groups and teams interact effectively during the life cycle of a project. This should not be used as an excuse for poor project performance, but as a catalyst to ensuring greater understanding of fundamental issues. Managers and participants alike need to respect the strength that diversity can bring to projects and put management processes in place that encourage effective and efficient interaction. Failing to take sufficient time to consider how people from different disciplines are likely to interact within a TPO will, most probably, lead to unnecessary difficulties and hence to inefficiencies and wasted effort.

It is easy to overlook the social context in which design and realisation activities take place. Project failure is often a result of a breakdown in communications, fuelled by cultural differences and divergent values. The greater the empathy between participants the closer they are in communication terms and the greater the potential for effective communication; conversely, the more distant they are in communication terms the greater the chance of ineffective communication. It is natural that businesses prefer to work with those they are familiar with and whom they can trust (based on previous project experience), in informal and formal alliances. However, because of requirements for competitive tendering on many projects the project relationships may be quite unexpected. Team assembly, maintenance and development stages should not be taken for granted as it is during this unpredictable and dynamic stage that many problems take root. The reluctance of the project manager and key stakeholders to deal with team assembly and maintenance may be related to:

- *Attitude.* Problems often relate to the project stakeholders having the right attitude to design and construction projects and the willingness to recognise the importance of social relations within projects. Failure to explore the attitude and values of key stakeholders before they are appointed may lead to problems as the project develops.
- *Inexperience.* Problems may be caused and/or go unnoticed if the main parties are inexperienced. TPOs should not be led by inexperienced project mangers without support from more experienced colleagues.
- *Ignorance.* The softer side of project management tends to be overlooked in many education programmes. Thus we have to learn through experience.
- *Failure to learn.* Failure to learn from previous project experiences is inexcusable but unfortunately too common as key actors rush from one project to another and priorities override common sense.
- *Priorities.* Even when the attitude is correctly aligned, knowledge of social relationships in projects is good and the ability to learn is exploited, the project priorities may make it difficult to apply some of the knowledge. Lack of time and finances are obvious factors and this relates to the attitude of the client.
- *People skills.* Not everyone is able to communicate and/or relate to others effectively, thus managers must be sensitive to personal traits.

Other challenges commonly experienced are:

- *Assuming too much.* We tend to assume that all actors are knowledgeable about the project, but the reality is that actors may have only a partial understanding of the entire project. Thus it is important to explore the boundaries of understanding throughout the project.
- *Failure to ask questions.* Assuming too much and concerns over professional reputation both combine to limit the amount and type of questions asked in face-to-face encounters.
- *Ineffective communication between groups.* This is often difficult to spot until there is a problem. Regular design reviews and meetings can help to identify some of the more obvious problems. All actors need to remain vigilant, especially the project and design managers.
- *Personality clashes.* These may be obvious from the start (and can be dealt with easily by substituting one or more individuals), but often personality clashes develop as people start working together and new actors enter and leave the TPO. It is impossible to prevent personalities from causing difficulties, but once evident they need to be quickly tackled. Failure to deal with differences may lead to ineffective communication and problems within the TPO.
- *Adversarial attitude.* Even within relational forms of contracting, such as project partnering and strategic alliances, there may be individuals who adopt an adversarial attitude to others. Sometimes this is evident from

the start of projects, but the attitude tends to develop or be revealed as the project proceeds and problems are encountered.

End of chapter exercises

- You are assigned to a project that is experiencing problems. The project manager is on sick leave because of occupational stress and is not likely to return in the near future. The project and the contributors are unfamiliar to you. What should your first task be?
- A situation has arisen that has caused a dispute between two individuals from separate organisations. The situation has the potential for deteriorating and affecting the performance of the TPO. If you were the project manager, what action would you take in an attempt to resolve the situation?
- Approximately halfway through the life of a major project you chair a progress meeting. During the course of the meeting it becomes apparent that the main contributors to the project are becoming jaded and the morale within the TPO is starting to decline. What measures should be taken to reinvigorate the TPO?

Further reading

For additional material on conflict try David Kolb's *Hidden Conflict in Organisations* (1992).

9 Learning

At the heart of a nation's AEC sector is the desire to improve the value delivered to clients, users and society. Innovations in both product and process are seen as the key to implementing change and improving our built environment. It is the ability to change and revise our (possibly outdated) ideas and practices that lies at the heart of process and product improvements. The large body of literature on diffusion of innovations (Rogers, 2003) has clearly demonstrated that the adoption of innovations – new ideas, practices and products – is not a given. The vast majority of innovations fail to gain a foothold, unable to dislodge established practices, or often take a long time to diffuse into widespread use. The reasons for the poor rate of uptake relate to the perceived advantages of the innovation compared with the established methods and also to the structure of the social system, be it an individual organisation or an entire discipline. As intimated throughout this book, the social structure of the TPO is dynamic and assembled from many different organisations, each with its own disposition toward innovations, thus it is unlikely that innovations will be adopted at a project level. Rather, the innovations tend to be adopted by one or more of the influential contributors and then applied to a project, with the result that the other contributors have little option but to adopt also.

Managing change is an important skill and a rich area for authors of change management literature. Overcoming resistance to organisational change has long been a concern for managers and management scientists. From around the late 1940s, literature has addressed resistance to change in organisations and a number of techniques and tools have been proposed (e.g. Lewin, 1951; DuBrin, 1974; Kotter and Schlesinger, 1979). Within this body of literature the implication is that change is always a good thing and that those who resist are some kind of organisational deviants, intent on undermining the organisation's productivity. Whether change is seen to be a good thing to be welcomed or a bad thing to be resisted depends on the individual, the culture of the organisation and the environmental conditions prevailing at the time, i.e. it depends on one's perspective and one's perceptions.

Learning involves a change in behaviour or understanding. If we are to change then we need to engage in learning and training activities. An essential competence of all project participants is the ability to systematically learn from projects. Evaluating the performance of projects (people and managerial systems) and products (people and technologies) can reveal a wealth of knowledge, most of which will be relevant to current projects and future work. Knowledge transfer between concurrent projects, and from completed project to new projects, is essential for improving the performance of project teams and the productivity of the organisations contributing to the projects. Knowledge gleaned from projects should be incorporated into office procedures to help keep operating procedures current, flexible and above all practical. To be successful the project and organisational cultures must be committed to a learning culture in which open communications and the ability to engage in constructive criticism are encouraged. A blame culture at the project and/or organisational level is not conducive to continual learning.

Fundamentals

Individuals should be engaged in continual learning and updating of skills through regular continual professional development activities, and these should be linked to better performance within the office and project environments. Evaluation and learning is a continual programme with long-term objectives for all stakeholders. Looking at how things are, watching how people behave, listening carefully to what people say and asking carefully framed questions are important skills. Similarly, the ability to assimilate knowledge from a variety of sources and present the findings to senior managers and user groups can be instrumental in helping to bring about positive change.

Measuring performance

Measuring performance is becoming increasingly widespread as organisations contributing to AEC projects aim to improve their performance through better management practices. A commitment by all members of the organisation to continual improvement is an excellent starting point; however, if performance is to be improved there must be procedures in place to report progress, measure it in a meaningful manner and then disseminate the results. Reward systems may also be introduced to encourage all members of an organisation to participate. For evaluation to be successful organisations and projects must have a clearly defined set of objectives. Performance targets need to be set out in the strategic and/or project brief, otherwise trying to gauge project success becomes a meaningless exercise.

To improve performance it is necessary to use some form of measurement. The mantra is 'if you cannot measure it, you cannot improve it'. The

Construction Industry Board produced the framework for key performance indicators in 1998. Working with the Department of Environment,
Transport and the Regions (DETR), Construction Best Practice Programme,
Construction Clients' Forum and the Movement for Innovation, data are
collected, collated and published. The indicators are designed to allow all
parties to the construction process to check how they are performing against
the industry as a whole. There are ten key performance indicators, seven
related to project performance and three related to the performance of the
organisation:

- project performance:
    - client satisfaction with the product (building)
    - client satisfaction with the construction process
    - defects
    - predictability of cost
    - predictability of time
    - actual cost
    - actual time
- company performance:
    - profitability – an important indicator for all organisations
    - productivity
    - safety.

Performance measurement is not just concerned with metrics; it is also
about the right attitude to continual improvement. Measuring needs to be
tailored to the size and scope of the organisation. Not enough data collection and analysis can result in a misleading picture and conversely too much
measuring can be wasteful and ultimately be self-defeating.

Lifelong learning

The philosophies of lifelong learning and continual professional development are fundamental to the continual development of individuals and their
organisations. Both professional and trade bodies require their members to
engage in, and provide evidence of, their commitment to continual improvement. There are a wide range of activities that constitute lifelong learning,
ranging from reading books and articles, to attending conferences and training programmes, through to undertaking a programme of study at a college
or university. The challenge for individuals and organisations is to integrate
the learning opportunities into the normal working week, i.e. making learning part of a balanced programme of work. This can be achieved through
the following:

- experiential learning from projects and the product
- reflection on work

- learning from how others approach their tasks
- learning from books and articles through evidence-based learning
- learning in action through action learning
- learning through storytelling.

It is well known that individuals have different learning styles (e.g. Kolb and Fry, 1975); therefore it is essential to understand how we best learn before embarking on a learning activity. This will help to maximise the limited time available and help to ensure a successful outcome. Evaluation and learning takes place on three levels, individual, organisational and project:

- *Individual needs.* Self-evaluation and learning is, perhaps, the easiest method, simply because it is in our own control. Engaging in reflective practice and undertaking formal (re)training courses may enhance our knowledge and skills. A relatively small amount of time is required for reflective practice, whereas training programmes may last for a few hours, half a day, a full day or even a few days, depending on the extent of the programme. Self-development may also be enhanced through research, for example undertaking a masters degree or a research programme (masters by research, MPhil, PhD). This involves a greater degree of commitment and resources.
- *Organisational needs.* Well-managed organisations have a comprehensive staff development plan and the resources to implement it. Organisational development will rely on a combination of individual self-development and formally organised group staff development activities in which specific staff will participate. Whether these sessions are optional or obligatory will depend upon the culture of the organisation and the importance of the subject matter. Investment in employee development schemes can help an organisation to stay competitive and respond to changing market conditions. As individuals grow their knowledge and skills the organisation collectively benefits.
- *Project needs.* Evaluation of projects spans organisational boundaries. Unless the activity is built into the project plan and facilitated by the project manager it is unlikely that learning from projects will be undertaken within the TPO. Instead, individuals and organisations will implement their own evaluation methods, usually retaining the knowledge gleaned for the event for their own employees' use. One way of trying to bring about learning across the entire TPO is to use meetings. Used sparingly, knowledge exchange meetings can form a platform for participants to share their knowledge with other participants with a view to bringing about improvements to the project performance.

Learning from our peers

Another approach to learning is through watching, listening and analysing the actions of our office colleagues and our fellow project participants. An enormous amount of knowledge can be gleaned, time permitting, simply by trying to make sense of what is going on in the work environment. Within the office interaction is relatively frequent and all-pervasive, and there are plenty of opportunities for learning if we can make the time to do so. Within the TPO interaction is less frequent, usually occurring in meetings and workshops, and so the opportunities to observe and listen to the actions of others are less frequent. An important competence of managers is the ability to look and listen to the activities of staff in the office and when possible the interaction with others involved in projects.

Newly qualified staff are usually allocated to an experienced member of the office so that they have some support in the early stages of their career. In these situations learning is intimate and the ability to ask questions and get immediate feedback helps to speed up the learning process.

Integrating research and practice

A number of schemes are available that seek to transfer knowledge between academia and industry. Knowledge Transfer Partnerships (KTPs) are one such vehicle for helping organisations to develop new ways of working, with the KTP associate acting as an interface between the university and the organisation participating in the KTP.

Other work-based programmes can be student-centred, with the majority of the work being undertaken in the workplace, or teacher-centred, with a structured programme of lectures, seminars and assessment. Many successful programmes combine both approaches and aim to develop the skills of staff around specific projects and/or work tasks. Work-based learning and development programmes seek to:

- identify shortfalls in existing knowledge and improve it
- identify new areas of knowledge and expertise
- encourage staff to share their knowledge (e.g. through internal seminars)
- better understand work practices with the aim of reducing waste and improving efficiency.

Informal arrangements can also be introduced, with organisations inviting academics into their offices to present and discuss ideas with them, possibly with a view to undertaking collaborative research projects.

Reflection in action

Experiential learning takes place through direct interaction with daily events, for example professional building designers working on a real life design project, and reflecting on those events (Kolb, 1984). The well-known experiential learning cycle (or learning spiral) contains four elements: concrete experience, observation and reflection, formation of abstract concepts and testing in new situations (Kolb and Fry, 1975).

More experience does not guarantee more learning. Learning from experience tends to be most effective when the experiences are painful or novel in some way, but learning from our mistakes is not a good policy (nor is it consistent with the total quality management ethos) if we wish to stay in business. The opportunity to learn from novel experiences may diminish as time passes. Thus reflective practice after the event is important because the individual will reflect on his or her actions, which may have been rather ordinary and uneventful, rather than waiting to learn from experience. The reflective practitioner has the opportunity to reflect on procedures and habits taken for granted, but which may be open to improvement when analysed. A number of tools to assist with reflective practice range from keeping a reflective journal to organised discussion groups with peers (quality circles). Reflection on practice is an essential component of professional development and the better the management of individuals' direction the better equipped the firm is to respond to change.

Dewey (1910) defined reflective thought as activities that are given persistent and careful consideration. The concept of the reflective practitioner (Schön, 1983, 1985, 1987) is well known to designers and forms an integral part of much architecture and built environment education. Individual reflection in action is a private activity, largely hidden from colleagues unless we decide to share the experiences. Reflection on daily events combined with evidence from published sources should form a systematic part of an individual's continual learning.

Reflective diary

A reflective diary or reflective log is an established tool for helping individuals to develop their knowledge and their ability to respond to situations. The intention is not to record every event in detail, but to record and reflect on events that are significant to an individual, the aim being that reflection puts the individual in a better situation should the same or a similar event occur in the future. The reflective diary may be a digital file (kept, for example, on a laptop) or a notebook or sketchbook, which tends to be preferred by designers because it is easier to add small schematic sketches. Frequency and style of entries is very much a personal choice. Some designers add entries to their diaries following a 'significant' event, for example a project suffering a major problem or conversely a major success. Others add entries on a weekly or even daily basis, recording and reflecting on less significant but equally important events associated with their job function. The recommended format is a simple three-stage approach in which individuals:

- *Describe the situation.* This should be a concise outline of the event, problem or success, the actors involved and the issue(s) to be reflected upon. Keep it factual.
- *Reflect on the event.* What could have been done differently? This tends to be personal.
- *Act.* Explore some scenarios. For example, how would you respond if faced with the same event in the future? What would you do differently? Is more knowledge, education or training required to help you be better prepared?

Reflective diaries are personal documents used by individuals to help improve performance. Indeed many find it useful and easier to reflect on work in the evening while at home. Diary contents are confidential; however, the process of reflection may result in the realisation that some issues cannot be tackled in isolation. These need to be raised within the organisation and/or within the TPO at an appropriate juncture. A reflective diary is also a good tool to identify things that went well and things that need to be improved. These issues can then be taken forward to regular knowledge exchange meetings and, if appropriate, the annual staff review.

Learning from projects

Learning opportunities are taking place throughout the life of a project. The challenge for individuals and managers alike is to incorporate specific events into project programmes in which knowledge can be exchanged and captured for use on current and future projects. Project control gateways are designed to discuss progress and hence engage in feedback related to specific activities. Similarly, formally constituted construction progress meetings provide another forum in which progress is discussed. Events at which the project team congregate may help to identify areas for improvement; however, the focus will be on the progress of the project and it is not uncommon for learning from the experience to be lost in the noise. For this reason it is recommended that specific learning events are built into project programmes at strategic intervals, where the focus will be on reflection and learning (not project progression). Learning feedback loops are crucial to the delivery of effective projects and to organisational learning. Similarly, the use of post-project reviews can help to identify good and bad experiences.

AEC professionals have often been criticised for not engaging in systematic project reviews, the feeling being that valuable knowledge is being lost between projects. This appears to be related to the constant pressure on professionals' time rather than any inability to do the review. It is the case that some organisations fail to build systematic project reviews into their working practices and project plans, but this is becoming less common. Failure to review projects may lead to the loss of important knowledge, resulting in the failure to identify and subsequently share good practice. This can lead to the replication of errors and missed opportunities to improve performance.

Reviews need to be facilitated, preferably by someone who has not been involved in the project and therefore lies outside the project culture. This can help actors to be a little more open and candid in their opinions. The facilitator is also more likely to ask some pertinent and perhaps unexpected questions of the project participants. On a practical level two project architects facilitating each other's projects can achieve this. Similarly, it may be useful for the design manager to facilitate such meetings. The outcome of the project reviews should be recorded in a concise report and sent to all those involved. All review activities need to clearly articulate the aim of the review and state those most likely to benefit from engaging in the event. The key objective may be to focus on team performance, cost control or control of design changes. Some design mangers will include reviews at strategic points in projects to focus on a topical issue. If actors are unwilling to engage in knowledge exchange, or are not candid, then it may be prudent to consider a tactical substitution for future projects.

Through project reviews

Through project reviews are normally incorporated into quality management systems and may be linked to routine project evaluation activities at control gateways. The benefit of conducting project reviews at key stages in the project is that there is less reliance on the memory of the actors than in post-project reviews. Knowledge harvested from such events can also be incorporated into other projects running concurrently, thus making the process more relevant to those involved. Care should be taken to clearly identify learning opportunities and separate this process from routine progress meetings. The advantage of through project reviews relates to the accessibility of key actors and current knowledge, assuming that the project culture is such that actors feel comfortable discussing areas for improvement. If the TPO is different each time it is unlikely that relationships will have developed to such an extent that the actors are willing to discuss issues openly, although this should improve as the project develops. Project team learning is essentially a collection of different learning experiences at group level, and this differs from learning in the more stable office environment. Through project reviews should:

- involve all key actors involved in the project at the time of the review; there are likely to be changes in some of the actors present between early and later reviews
- include the client

- clearly identify the topic that will be discussed
- disseminate findings (knowledge) to those present.

Post-project reviews

Post-project reviews (sometimes called project post-mortems) are an important component of any good quality management system. Evaluation at the end of the project aims to measure the success of the project against the goals set out in the brief. This exercise may generate valuable knowledge for inclusion in future projects, which will draw on and adapt the information, knowledge and experiences (both good and bad) generated during previous projects. This helps to keep organisations working efficiently while providing benefits for new projects. The members of the TPO mainly hold this knowledge; indeed team assembly is usually based on actors' previous project experience (building type, complexity, cost, performance, etc.) rather than professional discipline or educational qualifications. Post-project reviews are usually conducted at the end of a project or after the completion of a major phase of large projects. These meetings are often conducted by the project manager and/or design manager. The review will include input from major contributors to the project and is usually conducted as a meeting or a series of meetings. Participants are encouraged to share their experiences of what went well and what could have been done better so that future projects can benefit from the learning process. Some organisations may also conduct their own 'internal' post-project reviews to assist with organisational learning.

In situations in which post-project reviews are incorporated into the management plan for specific projects they may be an ineffective means of knowledge capture because:

- Some of the actors may have moved onto other projects and are unavailable for post-project data gathering. Many actors are only involved for a short period of time, to undertake a specific work package, after which they move to another project. This makes it very difficult to consult them after the event and helps to highlight the need for systematic project reviews during the life of the project.
- Pressures imposed by new projects may override those of completed projects, thus the review may be rather rushed and inconclusive, or may not happen, once again highlighting the need for reviews during the process.
- Actors may not remember all relevant facts, and if the review is conducted a long time after the start of a project it is highly likely that the opinions given are not as accurate or as informed as they might otherwise be.
- Professional rivalry may lead some actors to give a less candid account of the project, saving the 'real' knowledge for the benefit of their

organisation and their career advancement. This factor is relevant to all review activities.

- Professionals are reluctant to admit that they have made mistakes or have not performed as well as intended. Few of us feel comfortable discussing failures and mistakes openly and honestly unless there is an atmosphere of trust and mutual support (which is difficult to achieve within an organisation and even more so in a temporary grouping of project participants).
- Not all clients will be interested in post-project reviews. If the client is not a repeat client then they may have little to gain from taking part in a project post-mortem.
- Disagreements, disputes and conflict may make it impossible to conduct a post-project review.

Learning from products

Facilities management covers a wide range of activities, from space management and maintenance management to financial management, operations management and people management. Although definitions vary, the core function of the facilities manager is to support the core business activities of the organisation. The interface between facilities management and design is important if a building's potential is to be realised. The maintenance of a current set of plans and a maintenance database are paramount if a building is to be cared for and utilised effectively during its various lives. The challenge for all parties concerned with building maintenance, asset and facilities management is one of collecting dispersed, often inaccessible, information; this may lead to a loss of design and technical expertise that first created the building.

Of the vast quantity of information generated for individual projects much is related to the process of building (project-orientated information), with a small amount of information, such as drawings, specifications and maintenance logs, being applicable to the actual finished building (product-orientated information). Professionals involved in the building project will retain project information and product information for a set timescale, primarily for legal reasons. But this information is also a source of knowledge for new projects, and data retrieval and data mining computer software can greatly assist the design organisation's ability to retrieve past project information quickly. Project information may have little or no value to the owners/ occupiers of the completed building. Product information such as floor plans, details of construction and legal consents are needed by the building owners and its managers (not necessarily the client or original owners). Such information is easily stored and accessed electronically; however, many designers are reluctant to release this information fearing loss of future business (e.g. copyright laws and intellectual property rights). Increased interest in both building and service maintenance, coincident with the growth of the facilities management discipline, has brought a greater awareness of the value of accurate, accessible information.

Post-occupancy evaluation

There is an obvious link between the way in which we interact with buildings on a daily basis and how well the TPO performed. Collection and analysis of data from the building in use is termed post-occupancy evaluation (POE). Post-occupancy evaluation may be conducted by members of the original TPO and/or by consultants who have had no involvement in the design and realisation phases. Evaluation is usually undertaken at planned intervals after occupation, for example at twelve, twenty-four and sixty months. POE may be used to check that client and user values as stated in the written briefing documents were implemented as intended. The problem is that the users may be different from those represented in the early briefing exercises and this needs to be considered when analysing the data. The focus of the data collection may be coloured by the remit of the data collection exercise and the actors' experiences of the project. Key areas of concern related to building owner/occupier and building user interests are:

- *Space use.* Is the space functional and being used efficiently and as planned? How has the new facility impacted on working practices and productivity?
- *Time.* Has the new facility helped to improve flow of people within the building?
- *Well-being.* Are the perceived comfort and satisfaction of the staff acceptable? Has the internal environment helped with productivity?
- *Esteem.* How does the new facility impact on company image and what do the users feel about the building?
- *Energy use.* Is energy consumption as expected?
- *Maintenance and operating costs.* Replacement of components, cleaning, security, etc.

Collection of reliable data is much easier for some of these factors than others. For example, energy usage will be documented through meter readings, and smart systems can provide a wealth of information for analysis. In contrast, observing how people use space on a daily basis is more challenging. The intention of POE exercises is to report on existing performance and identify changes from the original brief. Recommendations for corrective action and a summary of the lessons learned can be made following analysis and reflection on the data collected. Addressing the issues should justify the expense of the action, leading to improved performance. Concomitant with other data collection exercises the purpose should be clearly defined and necessary approvals sought and granted from appropriate managers before data collection begins. Similarly, methodologies for evaluation should be kept simple, have measurable outcomes, be properly resourced and have a realistic time frame. Some of the data collection techniques used for POE studies are:

- *Observation*. Participant and non-participant observation can reveal rich data, but it is very time-consuming and building users may resist an outsider's presence. Few of us like to be watched as we work. Remote monitoring and surveillance through closed circuit cameras raises ethical issues and should be used only with the appropriate consents and safeguards in place.
- *Walk-through surveys*. Observational surveys are used to try and get an impression of how space is being used by looking at, and listening to, users. The term 'walk through' is a little misleading as data collection usually involves standing or sitting within spaces to observe users going about their daily work. Often used in conjunction with random informal discussions with users.
- *Questionnaires*. User satisfaction questionnaires can help to reveal perceptions of building users/owners/managers. Space usage quest-ionnaires are also used but may not give an accurate picture as they rely on people remembering when (and how) they used specific areas.
- *Interviews*. Interview techniques are useful for gathering user perceptions and opinions.
- *Focus groups*. Can be used to explore specific issues with a variety of stakeholders and user representatives.
- *Measurement*. Hard data collected through measurement are often easier to access and analyse: measuring space usage through electronic sensors, measuring energy usage, frequency of cleaning, replacement of short-life components, etc.
- *Benchmarking*. Allows comparisons to be made on a number of different levels, including with other buildings.

Evidence-based learning

Trying to make sense of daily challenges can be enhanced by comparing experience with relevant academic research findings and the views of others. Constant questioning can also help to keep knowledge fresh and relevant while stimulating innovative approaches to routine methods and proce-dures. Many professionals and tradespeople find it difficult to find the time to read research literature, often relying on the professional and trade jour-nals as a source of information and knowledge about the latest trends and innovations in their field. There are many other sources of knowledge that should be explored (time permitting), some of which may offer more value to practitioners than others. Typical sources include:

- *Textbooks.* Although textbooks are primarily aimed at students they represent a useful source of distilled knowledge that is readily accessible and, given the contents, cheap. Some of the better examples of textbooks tend to find their way onto professionals' bookshelves. Most publishers offer 'professional' titles that are aimed directly at professionals. These are sometimes presented in a handbook type format that the practitioner can consult during the working day.
- *Professional and trade journals.* These deal with topical issues and help to keep practitioners updated with developments in a range of areas, from design through to legal issues. They provide concise, easy to digest advice in a language accessible to the target reader.
- *Research journals.* A wide range of peer-reviewed journals are available within the AEC sector. They contain articles that have been peer reviewed by experts in the field, with the majority of articles requiring revisions, clarification and corrections before they can be accepted for publication. Peer-reviewed articles may offer practitioners some useful information; however, time will be required to search journals and databases for papers that are relevant to the interests of the individual and their organisation. Another challenge for practitioners is the language used in peer-reviewed journals, which is often difficult to access and understand.
- *Conferences.* The conference circuit tends to be occupied mainly by academics, many of whom have never practised, or seldom practise, their discipline. Thus practitioners may find the language used and the applicability of the espoused theory difficult to relate to their experience of daily practice. Some conference circuits have attempted to address this with 'industry' days or sessions built into the main conference to facilitate knowledge exchange between researchers and practitioners. Conferences can provide excellent networking opportunities as well as new knowledge on a particular subject. Conference proceedings vary in quality, ranging from papers that have not been peer reviewed through to proceedings in which the papers have been subject to extensive peer review before acceptance. As a general rule, if the ideas have some value it is highly likely that they will appear in a peer-reviewed journal sometime after the conference is held.
- *Continual professional development activities.* These are aimed at practitioners and tend to draw on information gleaned from conferences, journals and textbooks, usually enhanced with a fair amount of anecdotal evidence from active practitioners. Continual professional development offers a quick and effective means of improving and extending knowledge and skills, usually held for a day, half a day or a few hours in the evening.

As intimated above, one of the biggest challenges for the practitioner is finding the time to search out and read material relevant to his or her particular context and needs. For some people a good book may represent better value than attending a conference for a day; it depends on personal interests and needs, and the level of interaction sought with others. Whatever the approach taken, it is useful to look outside one's immediate field of experience and try and relate it to the experience of others working in other sectors. For example, industrial product design literature may be a fruitful source of information to readers looking for advice on design management. Similarly, readers involved in production may want to look to some of the mass production and bespoke production techniques used in, for example, the ship-building industry. Many of the ideas may need to be interpreted and modified slightly to suit the AEC context.

Action learning and action research

It is through reflection on individual and collective experience, combined with theoretical debates and analysis of research findings, that we are better able to implement improvements to our working practices. This involves planned (considered) change, much of which will be incremental and relatively gradual, some of which may be more radical and substantial. Two approaches, action learning and action research, may help groups to bring about change.

Action learning

'Action learning' is a term used to describe an inductive process in which managers seek to solve organisational problems within the workplace. This form of management development involves learning by the process of problem solving within a group, i.e. learning to learn by doing. As with action research, the problem needs to be clearly defined and the boundaries of the group determined before the process begins. Similarly the results of the study need to be analysed and disseminated within the organisation at the end of the programme. A facilitator is required to facilitate the progress of action learning.

Action learning is a collaborative learning method in which a small group of learners meet on a regular basis to address work issues. The philosophy is that effective learning will take place when we are faced with a real life problem to solve (Revans, 1998). The small group of learners is known as an 'action learning set', which draws on each individual's experience and knowledge to apply ideas to the work challenge. By working as a group the members tend to learn from others, applying knowledge and best practice to change their work. The practical outcome of the group exercise is the main outcome of the learning. This experiential and collective learning method can contribute to the development of individuals and organisations, bringing about changes in work practices and an increased capacity to learn.

There are a few essential rules of action learning that need to be followed. All group members are viewed as equals with a common aim; there is no leader or expert member of the group. Action learning is based on a real work challenge, for example a problem encountered on a current project. Group members are asked to reflect on their experience and their learning. To maintain equality within the group, and ensure that the group stays focused on the challenge, it is usually beneficial to facilitate the action learning set. An external facilitator will guide the group through the process, ensuring that equal participation rights are respected, that all members engage in reflection and that the enthusiasm to learn is sustained throughout the life of the action learning set.

Membership of the action learning set will be determined by the issue being addressed, be it an organisational challenge or a project-related issue. In medium to large organisations it may be possible to deal with the issue internally, drawing in group members from other departments. Alternatively, for smaller organisations it may be necessary or desirable to assemble the group from project contributors representing other organisations. Whatever the approach, the challenge must be clearly identified, a timescale agreed and the learning outcomes evaluated and disseminated as appropriate.

Action research

Action research is applied research that aims to actively and intentionally affect a change in a (social) system (Lewin, 1946). The research method involves a planned intervention by a researcher into naturally occurring events (Gummesson, 1991) and is a valuable variant of the quasi-experiment in management research (Gill and Johnson, 1997). This involves the active participation of the researcher(s) in the client system. For action research to have any value it must, like all other research, be conducted in a systematic manner, within a defined programme, and be adequately resourced. The success of this type of research will depend on the experience and competences of the researcher and the synergy between researcher and client. Success will also depend on the level of commitment shown by those taking part in the research. Ethical issues and the value of the research to both the client organisation and the researcher need to be discussed before entering into an agreement. The ethnographer may be someone with research training already working in the office (e.g. someone with a research degree) or a researcher may be invited into the client's organisation to carry out the research. The main stages in the research process follow a sequence similar to the following:

- *Start.* The client usually presents the problem, i.e. it is driven by the organisation, and discusses it with the researcher. Clear goals, resources and time frame are agreed. The amount of access to often confidential and commercially sensitive organisational settings also needs to be discussed and agreed.

- *Diagnosis*. This stage involves the researcher and client discussing and agreeing on the most appropriate management concepts and research tools to address the problem. An action plan is agreed.
- *Action*. The action stage involves the ethnographer collecting data from the workplace as the client implements the agreed action.
- *Evaluation*. Data are assembled and analysed jointly by the client and the researcher, the outcome of which should be recommendations and advice that can be taken forward and, if appropriate, developed into a new action plan. It is important to recognise that some action research may be inconclusive given the dynamic nature of design and construction projects. However, given the nature of the research it is unlikely that the effort will have been wasted as it will help to highlight how people act in the workplace.
- *Closure*. At the end of the programme generalisations may be made that help to inform business practices and it is likely that new problems and challenges will have emerged to be tackled in related research. Depending on the agreement between client and researcher the results may then be disseminated through publication, or retained for internal use only.

Storytelling

Storytelling can be a very useful way of helping to explain situations to others and can be an effective vehicle by which to transfer knowledge in organisations (e.g. Denning, 2001). Storytelling has also been used in architectural education as a way of exploring the 'black box' of design thinking (Heylighen *et al.*, 2007). Informal conversations and small group communication can be utilised to recount stories to new members of the office as a means of socialising then into the office norms, and to help illustrate good and bad practices. Similarly, newcomers to a project may benefit from listening to anecdotes to quickly familiarise themselves with commonly shared project values. Anecdotes sometimes become legends within the office, with the familiar opening line similar to 'Did you hear about the time when . . .' opening the way for a yarn relating to office and project morals, providing an indirect indication of the values held and expected within the organisation. No doubt the yarn involves a healthy mix of fact and fiction, indeed the truth is often stretched to make a point, but the aim is to get a message across that will be remembered. Stories are often used in health and safety inductions to make a point that will be remembered when visiting construction sites.

Stories are highly effective in helping to explain why things are done the way they are, and/or to make a point. Storytelling is an effective vehicle to transfer knowledge within or between small groups of people. Used effectively these conversations can help to expose and develop the knowledge held within offices and project organisations.

Practical challenges

Kaderlan (1991) noted that, in practice, individuals have only a partial understanding of and interest in what the organisation is doing; for example, architects will be primarily occupied with design issues, production staff will be concerned with production information, etc. Extending this thinking to the TPO, with the exception of the project managers, the various disciplines may have a limited understanding of the entire TPO. This can influence the manner in which individuals learn from their experiences, and it may be necessary for managers to implement a series of events to promote interaction and hence understanding of others' roles and responsibilities.

Failure to incorporate learning opportunities will usually result in a loss of valuable knowledge, not through any reluctance to analyse performance, but because other tasks become more pressing and the opportunities pass by. The main challenge to learning from both projects and products is related to forward planning and allocating sufficient time to the activity. With constant pressure to deliver faster and more efficiently it is easy to fall into a task-orientated mode with little time to reflect on the things that really matter; and thus fail to profit from the collective experience and knowledge. Some of the most common problems include:

- *Failure to plan.* This usually equates to a failure to learn.
- *Inappropriate attitude.* Learning is related to attitude and being able to keep an open and questioning mind. Sometimes the pressure of work can overwhelm even the most enthusiastic learner, and managers need to provide appropriate support to assist their staff in this regard.
- *Poor estimation* of the effort required to complete a task, poor programming and inadequate monitoring of progress. This usually results in individuals having to spend time allocated for other activities, e.g. continual professional development, on completing their work packages.
- *Poor communications within the office.* It is surprising how often individuals are unaware of their colleagues' actions, and thus fail to benefit from their experience and knowledge.

These challenges can be addressed by taking a strategic approach to learning, making the links between individual, organisational and project learning. Managers need to encourage and support their co-workers and ensure that learning events are built into all project programmes and form part of the organisation's culture. Providing an opportunity and encouraging staff to talk informally about the challenges they face in their daily tasks can help to disseminate knowledge and good practices within the organisation. Time for this important activity must be factored into the business overheads and incorporated into project programmes – otherwise it is highly likely that it will not happen. Systematic learning from the process and the product should form part of the office management system. Strategic incorporation of feedback from specific stages in the process is an essential component of a reflective and reflexive approach to management.

End of chapter exercises

- Assuming that you are involved in the early planning stages of a major project, how would you incorporate learning events into the project life cycle?
- 'Did you hear the one about . . .?' If you were a manager, what sort of story would you tell a new member of staff to introduce them to the health and safety culture of your current project?
- You discover a problem with ineffective communication between two of the organisations contributing to your current project. How could you learn from the situation? How would you disseminate what you learn to other members of the TPO?

Further reading

Readers interested in research methods applicable to their work setting should find *Research Methods for Managers* (Gill and Johnson, 1997) highly informative. For further information on experiential learning see *Experiential Learning* by D. A. Kolb (1984).

10 Implementation

In the preceding chapters some of the fundamental issues that influence the performance of interdisciplinary projects have been addressed. The underlying characteristics have deliberately been tackled in a generic way in order to provide insights, rather than prescribe 'how' to manage interdisciplinary TPOs. Although a generic approach has many benefits in an introductory text such as this, the underlying challenge remains one of implementation in a real life setting: we would all like to know if, and how, the theories work in practice. To shed a little light on the practical application of interdisciplinary management in practice a case study is used. This describes a people-centred approach to the management of AEC projects used by a large firm of consulting engineers in Denmark. This approach is not used universally within the organisation, which uses a variety of approaches to suit client's requirements on a project-by-project basis, but it has been used very successfully for a select number of projects.

The philosophy behind the approach is that value is the end goal of all construction projects and therefore the discussion and agreement of value parameters is fundamental to the achievement of improved productivity, client/user satisfaction and supply chain integration. The term 'value-based management' (and occasionally 'values-based management') has been used to describe their bespoke method for the creation and realisation of building designs. The method has its roots in supply chain logistics and lean thinking, using a structured approach to establish client values. Facilitated workshops are a central element of the method, used to bring key supply chain members together to interact, discuss and agree values. The workshops promote inter-personal communication and a team ethos throughout the project. Data for the case study were collected through a variety of methods, which included non-participant observation of the workshops, interviews with participants and analysis of project documentation. The chapter draws on longitudinal research, some of which has previously been reported by Emmitt and Christoffersen (2009) and Thyssen *et al.* (2010).

Context

The Danish construction sector, much like many others around the world, was criticised during the 1990s for its poor performance and failure to deliver value to the customer, resulting in government initiatives to promote improvements. Contrary to the criticism parallel research found that some innovative practices were being successfully implemented and improvements were being made by some organisations (Kristiansen *et al.*, 2005), which also reflects the situation in many other countries. As mentioned in the early chapters of this book, a great deal has been done by governments to promote change in their construction sectors, much of which has focused on the value delivered to the client while arguing for a more integrated, collaborative approach to the management of projects. Although this case study is taken from Denmark, a small country in Scandinavia, the ideas being implemented and the drivers behind the desire to improve the performance of TPOs are not dissimilar to initiatives taken in other parts of the world.

During the first decade of the twenty-first century a number of initiatives were taken by the case study organisation to (1) try and improve the value delivered to clients, users and society and (2) implement more efficient production processes. This included attention to technologies (e.g. manufacturing and ICTs) and also the management of construction activities. This helped to emphasise the importance of concepts such as lean construction, project partnering and more recently value management. Early trials included a consortium project representing the entire supply chain and the design and construction of student housing, which were completed within the time frame and to budget (see Emmitt and Christoffersen, 2009). These projects were also rated very highly by the client, project participants and end-users as measured in satisfaction surveys. More recently the model has been further developed in a number of consecutive projects within the greater Copenhagen area, including both new-build and refurbishment residential projects, a new biochemical plant, laboratory facilities and urban renewal of street lighting and paving for one of the municipal districts.

Underlying philosophy

The aim of the case study organisation was, and still is, to develop a better approach to the design and construction of buildings through attention to the entire supply chain. Central to this was the inclusion of the client in the TPO and a focus on value. Discussion and agreement of value parameters is fundamental to the achievement of improved productivity and client/user satisfaction. Value creation and value delivery are crucial components of the approach, explored and developed in what the case study organisation call the 'value universe' and implemented using a bespoke value-based design management model. This relies on the use of facilitated workshops and attention to the process of interaction during the life of the project.

Early models were based on approaching supply chain management from a logistics perspective (e.g. Christopher, 1998), followed by a focus on supply chain management from a lean production stance, drawing on the popular management literature of Womack *et al.* (1991) and Womack and Jones (1996). The seminal work of Koskela (1992, 2000) was also important in helping to emphasise the importance of processes as applied to AEC projects. This eventually resulted in the use of the Last Planner System (Ballard, 2000) for the construction phase, which was soon followed by the application of lean thinking to the client briefing and design phases. The process model has further evolved through a focus on value chains (Porter, 1985) and value management (e.g. Kelly and Male, 1993). Focusing on value has also led to increased interest in how project participants interact and communicate within a dynamic TPO (e.g. Emmitt and Gorse, 2003, 2007). Combined the ideas resulted in the development of a relatively simple process model, which is based on interaction within facilitated workshops throughout the entire life of the project. The main features of the model are based on:

- maximising value by means of a lean philosophy
- encouraging interaction and participation (including client involvement)
- transparent communication and trust throughout the entire supply chain.

A lean philosophy

The five principles of lean thinking (Womack *et al.*, 1991) have been interpreted and adapted to the design management phase of projects to be to:

- *Specify value*. Clearly and precisely identify the client's values and requirements, and then identify the specific functions required to deliver a solution.
- *Identify the value stream*. Identify the most appropriate processes to deliver the building through the integration of the functions identified when specifying value.
- *Enable value to flow*. Remove any unnecessary or redundant cost items from the design to get to the optimal solution (as agreed by the major project stakeholders).
- *Establish the 'pull' of value*. This means frequently listening to the client and other key stakeholders during the project and responding iteratively.
- *Pursue perfection*. Incorporate process improvement methods into the organisational culture and practices of the project participants' mother organisations.

These five principles underpin the workshop method, starting with the definition of value and continuing through the entire process, as described below. Lean thinking can be applied at different levels in the product development process, from the entire project to distinct phases and sub-stages, which can assist the planning and scheduling of the various work packages. Approaching design from a lean thinking perspective also helps to emphasise the need for designers to understand how design value is physically realised and the associated production costs, i.e. they need to understand the supply chain. Again this is addressed within the workshops. Depending on the type of project and the approach adopted by the design team this may involve a greater understanding of craft techniques or manufacturing production techniques, and the associated cost, time and quality parameters.

Interaction and participation

To explore values and implement lean thinking requires face-to-face inter-action within the TPO, and the primary mechanism for this is a series of facilitated workshops. This is not a new idea; for example the architects Konrad Wachsmann and Walter Gropius introduced a teamwork method for the development of complex building concepts in the 1940s (Gropius and Harkness, 1966). What has changed is that groupwork and teamwork have taken on more significance with the promotion of relational forms of contracting, integrated supply chain management and greater concern for client value.

Integration of supply chain members in the design (product development) process brings to the project the skills, knowledge and experience of a wide range of specialists, often working together as a virtual team from different physical locations. This requires social parity between actors, which means that professional arrogance, stereotypical views of professionals and issues of status have to be put to one side or confronted through the early discussion of values. To do this effectively all actors must engage in dialogue to explore and then confirm a set of values that form the basis of the project. The most effective way of doing this is through face-to-face meetings that recognise the value of group process (Luft, 1984). Interactions within groups, power relationships, leadership and decision making are extremely complex matters and contradictory views exist as to the ability of a group to reach its defined goals (e.g. Stroop, 1932; Yoshida *et al.*, 1978; Emmitt and Gorse, 2007). The philosophy is that unless interaction is addressed from the very start of projects in a professional and ordered manner it will be very difficult to achieve very high value.

The values-based process model

The literature on value management and value engineering overlaps, there-fore it is necessary to state how the two terms are interpreted and applied in relation to this case study. Value-based management attempts to control values, primarily through value management (see Kelly and Male, 1993) to 'create' value in the early stages of the project. Value-engineering techniques (see Miles, 1972) are used to 'deliver' value in the production stage. Values-based management uses face-to-face workshops as a forum to allow actors to discuss, explore, challenge, disagree and eventually agree to commonly shared project values. These values are then defined in a written document as a set of value parameters and prioritised in order of importance to the project team. This forms part of the project briefing (also known as architectural programming) documentation. Getting to know each other and thus establishing common values and/or knowing why values differ between the stakeholders is crucial to the method. It is about how to work together and how to keep agreements between the client and the delivery team.

In Denmark it is also common to differentiate between the values of the client (external values) and the values of the delivery team (internal values), and these are not to be confused (Christoffersen, 2003). External values are further separated into (1) process values and (2) product values. Process values comprise both 'soft' and 'hard' values. The soft values include work ethics, communication, conflict solving and trust between the client and the delivery team. These are intangible and difficult to measure objectively. The hard values include the delivery team's ability to keep to agreed time limits, cost estimates, quality of the product and workers' safety. These are tangible and can be measured objectively to assess project performance. Primary product values comprise beauty, functionality, durability, suitability for the site and community, sustainability and buildability. As the understanding of values improves and evolves through the design process, we are dealing with a learning process that relies on the development of trust and effective interpersonal communication.

Value design and value delivery

The process is separated into two main phases, the value design and the value delivery phases, which differ in their aims and management. Value design is where the client's wishes and requirements are identified and specified as a series of values through a value-mapping exercise. These values are then developed into a number of conceptual design alternatives. Management of the process should be focused on stimulating creativity and determining maximum value in the project, i.e. establishing needs before solutions.

Transition between the two phases is formalised in the construction contract. This is signed when, and only when, the value design work is complete, i.e. when everybody agrees that no more/no better value can come out of the project within the time available. Establishing and agreeing the 'point of no return', when the creative value design phase is replaced by the more pragmatic value delivery phase, is crucial for preventing rework and waste. The transformation point varies between projects and it is essential that all participants are aware of this shift in emphasis and respect it.

Value delivery is where the best design alternative, which maximises the client/customer value, is transformed through production into a finished artifact. The aim is to deliver the specified product in the best way and with minimum waste, using value chain mapping and value-engineering techniques. Value delivery comprises the final (detail) design and the construction of the project. Knowledge from contractors, sub-contractors and suppliers as well as knowledge and experience from using the building, or similar buildings, is incorporated by means of facilitated workshops. Management is concerned with keeping time, budgets and quality in a more traditional construction management context.

Interaction in workshops

Interaction in a facilitated workshop forum helps to establish common values and enables actors to better understand why their values differ. The creative workshops start with the agreement of common process values, followed by discussion of client intentions and abstract ideals. Then, work proceeds to produce a complete set of production information, before commencement of construction activities (see below). Workshops continue into the production phase, in which the main contractor gets the main sub-contractors involved. In fact, each workshop phase may comprise a series of facilitated workshops that deal with a particular issue, or value stage, which continue until agreement has been reached. Workshops are 'value generators' (or value drivers) and are concerned with problem framing.

It is a 'demand' of the values-based approach that the entire panel of participants is in place from the start of the project to its completion. For that reason the workshops tend to involve quite large numbers of people. Numbers present in the meetings vary between projects and stages, typically ranging between fifteen and thirty people. The organisational format of the workshops can be changed to accommodate more people if necessary by dividing into sub-groups. The number of workshops varies depending on the size and complexity of the project. Typically, workshops last for a half or a full day, but they never last for longer than one working day. Some flexibility in programming is required to accommodate the inherent uncertainty in knowing exactly how many workshops will be required to reach agreement. The experience of the facilitator is crucial here in accurately predicting the number of workshops necessary. When problems with understanding and attitudes persist, additional workshops are convened to explore the underlying values and tease out creative input. In extreme cases, if participants are unwilling to discuss and hence share values, they are asked to leave the process and are replaced by new participants. Experience has shown that incompatibility usually manifests in the first few workshops. Thus from the start of the project the whole process should be consensus-based and participants should have a shared vision and goals. The facilitator's role is to stimulate discussion, thus helping to identify areas of agreement and conflicting interests. It is important that he or she remains objective and neutral, allowing the participants to make the decisions within a facilitated and supportive environment, as discussed below.

A standard value agenda is used as a framework for decision making in the workshops. This is referred to as the 'basic value structure for buildings' and is based on the six product values (beauty, functionality, durability, suitability for the site and the community, sustainability, and buildability). This value hierarchy addresses the primary project objectives and breaks them down further into sub-objectives as part of an iterative process carried out within the workshops. Common value management tools, such as the value tree (see, for example, Dallas, 2006) and quality function deployment (QFD), are used to weight options (values) in a decision matrix to help find the solution that provides the best value. The process facilitator guides participants through the discussion of values in a systematic and objective way.

The workshop sequence in the value design phase

Workshop 0: (Partnering) Building effective relationships

The function of the preliminary workshop is to bring various actors together to engage in socialising and team-building activities. The intention is to build the communication structures for the project, thus allowing actors to engage in open and effective communication during the life of the project. In addition to setting the stage for the events that follow, the 'outcome' of the first workshop is the signing of a partnering agreement, which confirms the process values for cooperation. Early workshops are also concerned with the selection of the most appropriate consultants, based on their ability to contribute to the project (their 'fit') rather than the lowest fee bid. Collective dialogue helps to explore and develop relationships that can (or conversely cannot) develop into effective and efficient working alliances. Early workshops are also designed to help build a certain amount of trust and understanding before work commences.

Workshop 1: Vision

It is not possible to know values in depth at the start of a project, so workshops are primarily concerned with exploring values and establishing a common vision. Knowledge and experience from other projects are brought into the workshop, for example from facilities management. The main focus of the effort is the establishment of client values (value-based parameters), on the basis that the better these are known and clearly identified the better the team can deliver. An example could be functionality, sub-divided into optimal layouts to suit different users, for example office workers, visitors, cleaners and maintenance staff. Critical connections between decision making are explored so that everyone is certain of roles and responsibilities. The result of Workshop 1 is the establishment of basic values for the project, a very pragmatic document of prioritised values, which does not contain any drawings.

Workshop 2: Realism

Workshop 2 addresses how the basic project values may be fulfilled by presenting various design alternatives and looking at how they meet the basic value parameters. The contractual framework of the project is also addressed. Project economy is introduced here along with constraints associated with authorities, codes and regulations. Design proposals are worked through and ranked according to value. Architects are encouraged to produce at least three schemes that can be presented and discussed. Two to three workshops are normally required at this stage because there is much to discuss. Basic project values and project economy should be respected in this process and any changes justified within the value parameters. The outcome of the realism phase is the selection of the best-suited design proposal.

Workshop 3: Criticism

The presentation of the design proposals and the criticism is undertaken in two different workshops because it helps to encourage creativity and innovative solutions. In this workshop the proposed design solution is analysed and criticised to see if it really is the best solution and whether it could be improved. Discussion is centred on the chosen design solution and its potential for improvement within the agreed value parameters. There is usually some pressure at this stage to get the scheme into production quickly, and this has to be balanced against reducing uncertainty in the design before entering the production phases. Stakeholder satisfaction with the process value and the product value is measured on the basis of the partnering agreement and the basic product value parameters. This is done using key performance indicators at various stages in the process to measure the participants' perceived satisfaction. Then the project is approved for production and the contractual delivery specifications are fixed.

The workshop sequence in the value delivery phase

Workshop 4: Design planning

As the abstract design work turns into production information there is a shift in thinking. The value management techniques are supplemented with 'harder' value-engineering exercises. A process management tool is introduced to support process planning and define goals. This is currently based on a modified version of the Last Planner System (Ballard, 2000), which takes several issues into consideration and aims to give participants a clear view of what needs to be done, by whom and when. A design process plan maps activities and relationships and is an important coordination tool. Value-engineering and value-mapping exercises are conducted to identify and hence reduce waste. Many of the decisions are related to production activities, which are dealt with by interaction with the main contractor, working closely with the sub-contractors. Supply chain issues are planned and the first steps towards a production plan in the construction phase are taken in order to identify critical supply lines and accommodate any impact within the detailed design schedule.

Workshop 5: Buildability

Here the focus is on improving the buildability (or constructability) of the project, while trying to reduce waste in the detailed design and construction phases. The foremen and craftsmen meet with the designers to help discuss the efficient and safe realisation of the design. This often leads to some reconsideration of the design and revision of detailed designs to aid manufacturability and assembly. Changes may also be agreed to suit the available production capability and capacity. Once buildability issues have been resolved and agreed it is possible to move onto the final workshop stage in which the construction work is planned.

Workshop 6: Planning for execution

These workshops involve interaction between the main contractor and the sub-contractors to design and plan the construction process. A process plan is produced that helps to map the various production activities. This also helps to identify any missing or erroneous information. The Last Planner System is usually applied by the main contractor at this stage. Because of the high level of interaction during the earlier phases, many of the problems and uncertainties have been resolved, although the inclusion of sub-contractors allows for discussion about alternative ways of realising design value. On completion of the construction schedule, the information should be complete, thus providing the contractor and sub-contractors with a high degree of certainty in the production phases.

Leadership – the process facilitator's role

It is common for the client to employ the process facilitator directly to represent their interests. Alternatively, the contractor might pay for the facilitation role because the early resolution of problems and rapid development of trust within the team appears to be cost-effective over the course of a project, although this is difficult to prove in absolute terms. Regardless of who pays for the service, the process facilitator plays a key role in scheduling and facilitating the meetings. He or she usually has no contractual responsibilities and is not at liberty to contribute to the discussions, but is merely to try and ensure that all participants have equal participation rights. Thus the facilitator acts as an informal leader, charged with creating an effective social system that can drive the project forwards based on consensus. The responsibility of the facilitator extends only to the process, not the output of the process, which remains the responsibility of the team. The facilitator has no influence on the programme running alongside the workshops, other than to discuss and coordinate workshops with the project manager.

During the early meetings the facilitator is primarily concerned with creating a harmonious atmosphere within the workshops so that actors are able to communicate and share values, with the hope of reaching agreement. Negative conflict is managed to ensure that any disagreements are dealt with in a positive manner. Positive conflict and criticism are sometimes encouraged to try and prevent the manifestation of groupthink (Janis, 1972) and hence to try and keep the group from making poor decisions. The facilitator's role changes as the workshops proceed, with priority given to keeping the team together during difficult discussions in the later stages when cost and time tend to dominate the discussions (and when conflict is more likely to manifest). With no formal power the facilitator has to build trust and respect within the TPO to enable the workshops to function effectively. Moral support from formal managers, for example the project manager and the design manager, as well as the client is essential in this regard, helping the process facilitator to function as an effective informal leader. Needless to say, interpersonal communication between these parties must be effective and based on trust to allow the process to function. The process facilitator must possess excellent interpersonal skills and have sufficient knowledge of construction to be able to guide the process, allowing sufficient time for focused discussion on the task and time for socio-emotional interchanges that promote a team ethos. The aim is to encourage the formation and retention of interpersonal relationships.

Success of the facilitated workshops will be affected by the experience and skill of the facilitator. However, the actions of the participants are also a determining factor. Observations of meetings have revealed instances when participants have come to the meeting unprepared (e.g. cost information was not circulated before the meeting). This can cause a certain amount of turbulence, and sometimes this can result in the need for an additional workshop. In such situations the facilitator speaks to the 'problem' participant(s) outside the workshop environment to try and encourage better performance in future meetings. Observation has also revealed that a great deal of informal communication takes place before and after the formal workshop sessions.

Application

An illustration of the practical application of the workshop model is provided in this section. The project comprised two buildings of three and five floors respectively, with forty-two apartments and a total floor area of 3,600 m², located in the greater Copenhagen area. The apartments were designed for elderly people who required care and families with disabled children. Financial support was provided by the city council. The client owner is a non-profit organisation that owns 13,000 dwellings in Copenhagen. The term 'client' comprised a board of residents, which would be termed an experienced client in a Danish context.

Application of the workshop model was set up through earlier cooperation between the case study organisation, the client and the architectural firm and with financial support from the Danish Ministry of Social Welfare. Participants did not have any previous experience with the workshop model. The first workshop (Workshop 0) was conducted in the spring of 2005 and the process continued into the autumn of 2005. Observations of the early workshops indicated that the process was working well. However, at what turned out to be the last workshop in the autumn, major concerns over finances emerged that could not be resolved and subsequently the main contractor withdrew from the project. Two years later, in the autumn of 2007, a new contractor was found and it was decided to restart the project and the workshop process, both to integrate the contractor and to re-evaluate the original design, which was already well developed. Workshops were held with an average of thirteen participants comprising architects, engineers, a landscape architect, general contractor representatives, a process facilitator, a client project leader, a facility manager, a client in-house engineer and city representatives (some specialists did not attend all of the workshops).

The benefit of going through the workshop process again with a new contractor was new insights into the needs of the older end-users, who were less mobile and needed more care than originally foreseen. Representatives of the city council provided this information, which initiated a very creative design process to create more space in the bathrooms and bedrooms and improve the indoor climate – all which were to be accomplished within the budget. This was achieved mainly by removing some kitchen and basement facilities that the older users would rarely use, as well as making some changes to the windows and removing some built-in wardrobes. Subsequently a cheaper facing brick was chosen to keep the project within budget. In this process all participants contributed; however, it was critical that the contractor's representatives were experienced enough to make cost estimates on the spot (which had not been the case in the first application of the workshop model, which had hampered the decision-making process). The client expressed great satisfaction with the changes and the contractor found it motivating to know that the facility would satisfy the needs of the end-users. Not everyone was satisfied, with the architects expressing some annoyance about making the design changes and the additional work that this involved.

Some other changes were suggested by the client, which were found not to be feasible. However, additional assessments were made to ensure that changes could be made after the building was completed if they were found to be essential and if additional finances were available. In this way adaptability was considered. It should also be noted that, within this process, it seemed to be a supporting factor that the client representative was enthusiastic about the concept of the workshop model and possessed the authority of an experienced professional who was able to make decisive decisions; this had a contagious effect on the rest of the TPO.

There were, however, a few challenges that emerged. Several comments were made by the participants regarding the workshop model being very time-consuming (each session lasted about four to five hours), a point discussed in more detail below. After Workshop 3 the architect and contractor were confident about the way ahead and were so eager to proceed with the detailed design and construction that they did not consider a fourth workshop, about integrated design and construction scheduling, to be necessary; so they carried on without it. Thus, in this example, the process was not fully followed. It should also be noted that the participants were much better at discussing technical solutions than addressing value or values.

Discussion

Even though the workshop model had already been conducted to quite an advanced stage, a second round of discussing client needs with a new contractor gave way for new insights and an improvement of 'product value' from a user perspective. This highlights the importance of taking the time to understand client needs and corresponds with the notion that needs (value) change over time. Analysis of the data indicated that the workshop model can facilitate good cooperation and decision making through discussion of 'process values'. Furthermore, the case study highlighted the importance of having experienced participants with decision-making authority, who can engage constructively and make decisions without having to refer back to their superiors. This calls for a thorough analysis of the participants before making invitations to the workshops.

Professionals' time is valuable and great effort should be directed towards limiting the duration of the workshops. Also, clear agreement should be made about payment for making more than one design alternative and the subsequent changes to design caused by the workshop approach. Greater acceptance of these changes may also be facilitated by some negotiation of the client values with the delivery team to establish commitment. Discussing value is difficult and a more rigorous use of quality function deployment may help to translate the client values into a language understandable to building professionals. Alternatively, one should accept that value may not be stated easily and instead of trying to 'freeze' client value statements, the value tree should be perceived as a dynamic document reflecting an ongoing value conversation/interpretation.

At the end of Workshop 3 an evaluation of the process was conducted among the participants by means of anonymous questionnaires. This resulted in the participants reporting a highly positive experience, reinforcing earlier research by the Danish Building Research Institute, which found improved performance across a whole range of performance parameters when applying the values-based model (By og Byg, 2004; SBi, 2005). In addition to this, participants have consistently evaluated the process highly, finding it an enjoyable and productive way of working. Further work is required, however, to investigate the effectiveness of the workshop method in terms of the realisation of group goals. The literature on effective groups and teams suggests that the group size should be just large enough to include individuals with the relevant skills and knowledge to solve the problem; this is the principle of least group size (Thelen, 1949). The optimum size is considered to be around five or six people (Hare, 1976), which is considerably smaller than the composition of the workshops. Other areas of potential research relate to the skills and competences of the process facilitator, not just in facilitating the meetings but also as a socialising function of project management. Some investigation of interpersonal communication skills (task-based and socio-emotional) may also be useful avenues to explore in terms of educating/training process facilitators.

Reflection

Bringing people together in facilitated workshops is time-consuming and given the number of experts involved it is expensive. However, the workshops have proven to be an essential forum for discussing differences, bringing about good design decisions and resulting in agreement being reached. Workshops have also helped to encourage open communication and knowledge sharing, with learning as a group contributing to the clarification and confirmation of project values. It appears that value is not something that can be made explicit once and for all by writing it down in a fixed value system for subsequent design evaluation, which is common practice in value management (Kelly *et al.*, 2004); instead value is continually changing as the project evolves.

The number of workshops required to ensure that all participants reach agreement on the project value parameters (or at least establish areas in which consensus is not reached and why) can sometimes exceed that planned, which can pose challenges for the scheduling of work. For projects with very tight schedules such uncertainty can present problems for project management teams that are not familiar with the approach. Experienced facilitators are able to bring people together quickly and usually conduct the workshops efficiently. Sometimes it is simply a matter of discussing and agreeing on whether or not time is the crucial constraint for the project. The schedule of meetings may be extensive on a large project and there is a concern that the cost of the meetings may outweigh the value realised through them. There is also the constant danger of holding too many workshops and the participants becoming jaded though overfamiliarisation. All participants need to constantly monitor the effectiveness of the workshops and critically assess their added value through the use of various benchmarking tools. Although the workshops act as informal control gates, there are no formal gates (unlike some other process models). Some consideration of more formal procedures in line with total quality management could help the process facilitator and project managers to coordinate programmes a little better.

Conclusion

The way in which people interact within the project environment and with their colleagues in their respective organisations will have a major influence on the success of individual projects and the profitability of the participating organisations. Both the metaphorical and physical space between the organisations participating in projects will influence interaction practices and hence the effectiveness of the project outcomes. This means that (considerable) effort is required in trying to manage interpersonal relationships to the benefit of clients' projects and also to the profitability of the organisations contributing to them. Emphasis should be on maximising value through improved interaction, communication and learning within the entire supply chain.

Although the approach described in this chapter does work extremely well for some clients in a Danish (democratic) context, it might not suit all clients or all societies. There have also been some instances in which the facilitated workshop method has been applied late in the process to projects that are experiencing problems, and these 'insertions' have proven to be ineffective despite the efforts of the facilitators (as the problems have already manifested). The approach works only when all participants engage in open communication, and this takes a shift in thinking for many of the participants who are more familiar with adversarial practices and closed (defensive) communication. However, the underlying principles should be of interest to a wide range of disciplines involved in the management of successful AEC projects.

Exploring values is complicated and trying to satisfy all stakeholders is rather ambitious. It is also important to emphasise that the approach described in this chapter has evolved over a number of years and is one of many approaches to try and improve the Danish AEC sector. As intimated above, it is a challenge to implement the values-based model in practice, requiring considerable effort and determination on behalf of the process facilitator and commitment from the project participants.

From a research perspective a major concern is that practical improvements to the workshop model will be difficult to measure in a rigorous manner because of the technical uniqueness and contextual difference (social construction) of each AEC project, which makes it impossible to establish a norm. Face-to-face interviews, as a supplement to observations, can provide some accounts of the effects of the workshop model from the perspective of the project participants, who can relate their experience to their 'normal' practice. The subjective concept of 'satisfaction' or 'value' can be dealt with through interviews, even though interview accounts have many kinds of bias. These challenges are not unique, but they highlight some of the difficulties of reporting practical work in the AEC field. Further applied research into the composition of TPOs, their collective performance over the life of projects and the nature of interpersonal communication practices must be undertaken in live projects to investigate how different communication patterns affect group, and hence project and organisational, performance (Emmitt and Gorse, 2007).

End of chapter exercises

- Assuming that you were asked to plan a new project using the workshop approach described in this chapter, how would you plan the process and schedule the workshops (assuming a twenty-four-month timescale from inception to completion)?
- How would you introduce the main project stakeholders to the workshop method?
- If one of the participants decided that they did not want to participate in the workshops, what would you do?

References

Abrahamson, E. (1996) 'Management fashion', *Academy of Management Review*, **21** (1), 254–85.

Ackoff, R. L. (1966) 'Structural conflict within organisations', in J. R. Lawrence (Ed.), *Operational Research and the Social Sciences*, Tavistock Publications, London.

Allport, G., Vernon, P. and Lindsey, G. (1960) *A Study of Values* (third edition), Houghton Miflin, Boston, MA.

Anderson, C. M., Riddle, B. L. and Martin, M. M. (1999) 'Socialisation processes in groups', in L. R. Frey (Ed.), *The Handbook of Group Communication Theory and Research*, Sage Publications, London, pp. 139–63.

Arnstein, S. R. (1969) 'A ladder of citizen participation', *Journal of American Institute of Planners*, **35** (4), 216–24.

Atkinson, S. and Butcher, D. (2003) 'Trust in managerial relationships', *Journal of Managerial Psychology*, **18** (4), 282–304.

Austin, S., Baldwin, A., Hammond, J., Murray, M., Root, D., Thomson, D. and Thorpe, A. (2001) *Design Chains: a handbook for integrated collaborative design*, Thomas Telford, Tonbridge.

Baden Hellard, R. (1995) *Project Partnering: principle and practice*, Thomas Telford, London.

Bales, R. F. (1950) *Interaction Process Analysis: a method for the study of small groups*, Addison-Wesley Press, Cambridge, MA.

Bales, R. F. (1970) *Personality and Interpersonal Behaviour*, Holt, Rinehart and Winston, New York.

Ballard, G. (2000) *The Last Planner System*, PhD thesis, University of Birmingham, UK.

Belbin, R. M. (1981) *Management Teams: why they succeed or fail*, Heinemann, Oxford.

Belbin, R. M. (1993) *Team Roles at Work*, Butterworth-Heinemann, Oxford.

Belbin, R. M. (2000) *Beyond the Team*, Butterworth-Heinemann, Oxford.

Bemm, D. J., Wallach, M. A. and Kogan, N. (1970) 'Group decision making under risk of aversive consequences', in P. Smith (Ed.), *Group Processes Selected Readings*, Penguin, Middlesex, pp. 352–66.

Bennett, J. and Jayes, S. (1995) *Trusting the Team: the best practice guide to partnering in construction*, Thomas Telford, Tonbridge.

Blau, P. M. (1964) *Exchange and Power in Social Life*, Wiley, New York.

Blyth, A. and Worthington, J. (2010) *Managing the Brief for Better Design* (second edition), Spon Press, London.

Boyd, D. and Chinyio, E. (2006) *Understanding the Construction Client*, Blackwell Publishing, Oxford.

Brenkert, G. (1998) 'Trust, business and business ethics: an introduction', *Business Ethics Quarterly*, **8** (2), 195–203.

Brensen, M., Goussevskaia, A. and Swan, J. (2005) 'Implementing change in construction project organisations: exploring the interplay between structure and agency', *Building Research & Information*, **33** (6), 547–60.

Brown, R. (2000) *Group Processes: dynamics within and between groups* (second edition), Blackwell, Oxford.

Brownell, H., Pincus, D., Blum, A., Rehak, A. and Winner, E. (1997) 'The effects of right hemisphere brain damage on patients: use of terms of personal reference', *Brain and Language*, **57**, 60–79.

Brynjolfsson, E. (1993) 'The productivity paradox of information technology: review and assessment', *Communications of the ACM*, **36** (12), 66–78.

By og Byg (2004) 'Evaluering af forsøg med trimmet projektering og trimmet byggeri', Report number 421–047, January 2004, By og Byg, Statens Byggeforskningsinstitut, Hørsholm.

Campbell, A. C. (1968) 'Selectivity in problem solving', *American Journal of Psychology*, **81**, 543–50.

Cartwright, D. and Zander, A. (Eds) (1968) *Group Dynamics: research and theory* (third edition), Harper & Row, New York.

Caudill, W. W. (1971) *Architecture by Team: a new concept for the practice of architecture*, Van Nostrand Reinhold, New York.

CIOB (2006) *Corruption in the UK Construction Industry – Survey 2006*, Chartered Institute of Building, Ascot.

Christoffersen, A. K. (2003) *State of the Art Report: working group value management*, Byggeriets Evaluerings Center, Copenhagen.

Christopher, M. (1998) *Logistics and Supply Chain Management: strategies for reducing costs and improving services* (second edition), Pitman, London.

Cline, R. J. W. (1994) 'Groupthink and the Watergate cover-up: the illusion of unanimity', in L. R. Frey (Ed.), *Group Communication in Context: studies of natural groups*, Lawrence Erlbaum Associates, Mahwah, NJ, pp. 199–223.

Collaros, P. A. and Anderson, L. R. (1969) 'Effect of perceived expertness upon creativity of members of brain storming groups', *Journal of Applied Psychology*, **53** (2), 159–63.

Cuff, D. (1991) *Architecture: the story of practice*, MIT Press, Cambridge, MA.

Dainty, A., Moore, D. and Murray, M. (2006) *Communication in Construction: theory and practice*, Taylor & Francis, London.

Dallas, M. F. (2006) *Value and Risk Management: a guide to best practice*, Blackwell Publishing, Oxford.

Damodaran, L. and Shelbourn, M. (2006) 'Collaborative working – the elusive vision', *Architectural Engineering and Design Management*, **2** (4), 227–43.

Denning, S. (2001) *The Springboard*, Butterworth-Heinemann, Boston.

Dewey, J. (1910) *How We Think*, D.C. Heath & Co., Cambridge, MA.

DuBrin, A. (1974) *Fundamentals of Organisational Behaviour: an applied perspective*, Pergamon, New York.

Egan, J. (1998) *Rethinking Construction: the report of the construction task force*, DETR, London.

Egan, J. (2002) *Rethinking Construction: accelerating change*, Strategic Forum for Construction, London.

Emmerson Report (1962) *Survey of the Problems before the Construction Industry*, HMSO, London.

Emmitt, S. (1999) *Architectural Management: a competitive approach*, Longman, Harlow.

Emmitt, S. (2007) *Design Management for Architects*, Blackwell Publishing, Oxford.

Emmitt, S. and Christoffersen, A. K. (2009) 'Collaboration and communication in the design chain: a value-based approach', in W. J. O'Brien, C. T. Formoso, R. Vrijhoef and K. A. London (Eds), *Construction Supply Chain Management Handbook*, CRC Press, Boca Raton, FL, section 5, pp. 1–17.

Emmitt, S. and Gorse, C. A. (2003) *Construction Communication*, Blackwell Science, Oxford.

Emmitt, S. and Gorse, C. A. (2007) *Communication in Construction Teams*, Spon Research, Taylor & Francis, Oxford.

Emmitt, S. and Yeomans, D. T. (2008) *Specifying Buildings: a design management perspective* (second edition), Butterworth-Heinemann, Oxford.

Emmitt, S., Prins, M. and Otter, A. den (2009) *Architectural Management*: *international research and practice*, Wiley-Blackwell, Oxford.

Fiske, J. (1990) *Introduction to Communication Studies* (second edition), Routledge, London.

Feldman, D. C. (1981) 'The multiple socialization of organization members', *Academic Management Review*, 9, 47–53.

Forsyth, D. R. (2006) *Group Dynamics* (fourth edition), Thomson Wadsworth, Belmont, CA.

Freudenberger, H. J. (1974) 'Staff burnout', *Journal of Social Studies*, 30, 159–65.

Fukuyama, F. (1995) *Trust: the social virtues and the creation of prosperity*, Hamish Hamilton, London.

Furnham, A. (1986) 'Situational determinants of intergroup communication', in W. B. Gudykunst (Ed.), *Intergroup Communication: the social psychology of language*, Edward Arnold Publishers, Baltimore, pp. 96–112.

Gameson, R. N. (1992) *An Investigation into the Interaction between Potential Building Clients and Construction Professionals*, PhD thesis, University of Reading.

Gameson, R. N., Sherrat, R. D. S. and Bellamy, T. (2005) *Generic Skills in Design Teams: final report*, University of Newcastle, Australia.

Giles, H. (1986) 'General preface', in W. B. Gudykunst (Ed.), *Intergroup Communication: the social psychology of language*, Edward Arnold Publishers, Baltimore.

Gill, J. and Johnson, P. (1997) *Research Methods for Managers* (second edition), Paul Chapman, London.

Gorse, C. A (2002) *Effective Interpersonal Communication and Group Interaction During Construction Management and Design Team Meetings*, PhD thesis, University of Leicester.

Gorse, C. and Emmitt, S. (2009) 'Informal interaction in construction progress meetings', *Construction Management & Economics*, 27 (10), 983–93 (special issue, 'Informality and Emergence in Construction', edited by Paul Chan and Christine Räisänen).

Gouran, D. S. and Hirokawa, R. Y. (1996) 'Function theory and communication in decision-making and problem solving groups: an expanded view', in R. Y. Hirokawa and M. S. Poole (Eds), *Communication and Group Decision Making* (second edition), Sage, Thousand Oaks, CA, pp. 55–80.

Gray, C and Hughes, W. (2001) *Building Design Management*, Butterworth-Heinemann, Oxford.

Grilo, L., Melhado, S., Silva, S. A. R., Edwards, P. and Hardcastle, C. (2007) 'International building design management and project performance: case study in Sao Paulo, Brazil', *Architectural Engineering and Design Management*, 3 (1), 5–16 (special issue, 'Aspects of Building Design Management', edited by S. Emmitt).

Gropius, W. and Harkness, S. P. (Eds) (1966) *The Architects' Collaborative, 1945–1965*, Tiranti, London.

Gummesson, E. (1991) *Qualitative Methods in Management Research*, Sage, London.

Gutman, R. (1988) *Architectural Practice: a critical view*, Princeton Architectural Press, Princeton, NJ.

Gutzmer, W. E. and Hill, W. F. (1973) 'Evaluation of the effectiveness of learning through the discussion method', *Small Group Behaviour*, 4 (1), February, 5–34.

Hackman, J. R. (1992) 'Group influences on individuals in organisations', in M. D. Dunnette and L. M. Hough (Eds), *Handbook of Industrial and Organisational Psychology* (second edition), Consulting Psychologists Press, Palo Alto, CA, pp. 199–267.

Handy, C. B. (1995) *Beyond Certainty: the changing worlds of organisations*, Hutchinson, London.

Hare, A. P. (1976) *Handbook of Small Group Research* (second edition), The Free Press, New York.

Hargie, O. D. W., Dicksos, D. and Tourish, D. (1999) *Communication in Management*, Gower, Hampshire.

Hartley, P. (1997) *Group Communication*, Routledge, London.

Hassinger, E. (1959) 'Stages in the adoption process', *Rural Sociology*, 24, 52–3.

Heinicke, C. and Bales, R. F. (1953) 'Development trends in small groups', *Sociometry*, 16, 7–38.

Heylighen, A., Martin, W. M. and Cavallin, H. (2007) 'Building stories revisited: unlocking the knowledge capital of architectural practice', *Architectural Engineering and Design Management*, 3 (1), 65–74 (special issue, 'Aspects of Building Design Management', edited by S. Emmitt).

Hoffman, J. and Arsenian, J. (1965) 'An example of some models applied to group structures and processes', *International Journal of Group Psychotherapy*, 15, 131–53.

Hollingshead, A. B. (1998) 'Communication, learning and retrieval in transactive memory systems', *Journal of Experimental Psychology*, 34, 423–42.

Hosking, D. and Haslam, P. (1997) 'Managing to relate: organizing as a social process', *Career Development International*, 2 (2), 85–89.

Jackson, J. (1965) 'Social stratification, social norms, and roles', in I. D. Steiner and M. Fishbein (Eds), *Current Studies in Social Psychology*, Holt, Rinehart and Winston, New York, pp. 301–9.

Janis, I. L. (1972) *Victims of Groupthink: a psychological study of foreign policy decisions and fiascos*, Houghton Mifflin, Boston, MA.

Jarvenpaa, S. L. and Leidner, D. E. (1999) 'Communication and trust in global virtual teams', *Organization Science*, 10 (6), 791–815.

Jørgensen, B. and Emmitt, S. (2008) 'Lost in transition – the transfer of lean manufacturing to construction', *Engineering Construction & Architectural Management*, **15** (4), 383–98.

Kaderlan, N. (1991) *Designing Your Practice: a principal's guide to creating and managing a design practice*, McGraw-Hill, New York.

Karl, K. A. (2000) 'Trust and betrayal in the workplace: building effective relationships in your organisation', *Academy of Management Executive*, **14**, 133–5.

Keller, R. T. (2001) 'Cross-functional project groups in research and new product development: diversity, communications, job stress, and outcomes', *Academy of Management Journal*, **44**, 547–55.

Kelly, J. and Male, S. (1993) *Value Management in Design and Construction: the economic management of projects*, E & FN Spon, London.

Kelly, J., Male, S. and Graham, D. (2004) *Value Management of Construction Projects*, Blackwell, Oxford.

Keyton, J. (1999) 'Relational communication in groups', in L. R. Frey (Ed.), *The Handbook of Group Communication Theory and Research*, Sage, London, pp. 192–221.

Kolb, D. (1992) *Hidden Conflict in Organisations*, Sage Publications, London.

Kolb, D. A. (1984) *Experiential Learning*, Prentice-Hall, Englewood Cliffs, NJ.

Kolb, D. A. and Fry, R. (1975) *Theories of Group Processes*, John Wiley & Sons, Chichester.

Korsgaard, C. M. (1986) 'Aristotle and Kant on the source of value', *Ethics*, **96** (3), April, 486–505.

Koskela, L. (1992) *Application of the New Production Philosophy to Construction*, Centre for Integrated Facility Engineering, Stanford, CA.

Koskela, L. (2000) *An Exploration towards a Production Theory and its Application to Construction*, VTT, Espoo, Finland.

Kotter, J. P. and Schlesinger, L. A. (1979) 'Choosing strategy for change', *Harvard Business Review*, March/April, 106–14.

Kramer, R. M. (1999) 'Trust and distrust in organizations: emerging perspectives, enduring questions', *Annual Review of Psychology*, **50**, 537–67.

Kristiansen, K., Emmitt, S. and Bonke, S. (2005) 'Changes in the Danish construction sector: the need for a new focus', *Engineering Construction and Architectural Management*, **12** (5), September/October, 502–11.

Latham, M. (1993) *Trust and Money: interim report of the joint government industry review of procurement and contractual arrangements in the United Kingdom construction industry*, HMSO, London.

Latham, M. (1994) *Constructing the Team: final report*, HMSO, London.

Lamm, H. and Trommsdorff, G. (1973) 'Group versus individual performance tasks requiring ideational proficiency: a review', *European Journal of Social Psychology*, **3**, 361–88.

LeDoux, J. (1998) *The Emotional Brain*, Phoenix, New York.

Lee, F. (1997) 'When the going gets tough, do the tough ask for help? Help-seeking and power motivation in organisations', *Organisational Behaviour and Human Decision Processes*, **72** (3), 336–63.

Lewin, K. (1946) 'Action research and minority problems', *Journal of Social Issues*, **2**, 34–6.

Lewin, K. (1951) *Field Theory in Social Science*, Harper & Row, New York.

Lieberman, M., Lakin, M. and Whitaker, D. (1969) 'Problems and potential of psycho-analytic and group theories for group psychotherapy', *International Journal of Group Psychotherapy*, 19, 131–41.

Littlepage, G. E. and Silbiger, H. (1992) 'Recognition of expertise in decision-making groups: effects of group size and participation patterns', *Small Group Research*, 22, 344–55.

Love, P. E. D., Irani, Z., Li, H. and Cheng, E. (2001) 'An empirical analysis of IT/IS evaluation in construction', *Journal of Construction Information Technology*, 8 (1), 15–27.

Luft, J. (1984) *Group Process: an introduction to group dynamics*, Mayfield, Palo Alto, CA.

Mabry, E. A. (1985) 'The systems metaphor in group communication', in L. R. Frey (Ed.), *The Handbook of Group Communication Theory and Research*, Sage, London, pp. 71–91.

March, J. G. (1994) *A Primer on Decision Making: how decisions happen*, Free Press, New York.

Maslach, C. (2003) 'Job burnout: new directions in research and intervention', *Current Directions in Psychological Science*, 12 (5), 189–92.

Matthews, R. A. J. (1997) 'The science of Murphy's Law', *Scientific American*, April, 276 (44), 88–91.

Mayer, R. C., Davis, J. H. and Schoorman, F. D. (1995) 'An integration model of organizational trust', *Academy of Management Review* 20 (3), 709–34.

Maister, D. (1993) *Managing the Professional Service Firm*, The Free Press, New York.

McCroskey, J. C. (1997) 'Willingness to communicate, communication apprehension, and self-perceived communication competence: conceptualizations and perspectives', in J. A. Daly, J. C. McCroskey, J. Ayres, T. Hopf and D. M. Ayes (Eds), *Avoiding Communication: shyness, reticence and communication apprehension*, Hampton Press, Cresskill, NJ, pp. 75–108.

McKnight, D. H., Cummings, L. L. and Chervany, N. L. (1998) 'Initial trust formation in new organizational relationships', *Academy of Management Review*, 23, 473–90.

Mehrabian, A. (2007) *Nonverbal Communication*, Transaction Publishers, New Brunswick.

Menger, C. (1950) *Principles of Economics* (Trans. Dingwall, J. and Hoselitz, B. F.), The Free Press, New York.

Meyerson, D., Weick, K. E. and Kramer, R. M. (1996) 'Swift trust and temporary groups', in R. M. Kramer and R. T. Tyler (Eds), *Trust in Organisations: frontiers of theory and research*, Sage Publications, Thousand Oaks, CA, pp. 166–95.

Miles, L. D. (1972) *Techniques of Value Analysis and Engineering*, McGraw Hill Higher Education, New York.

Mintzberg, H. (1973) *The Nature of Managerial Work*, Harper and Row, New York.

Morgan, B. B. and Bowers, C. A. (1995) 'Teamwork stress: implications for team decision making', in R. A. Guzzo and E. Salas (Eds), *Team Effectiveness and Decision Making in Organisations*, Jossey-Bass, San Francisco, pp. 262–90.

Morgan, J. M. and Liker, J. K. (2006) *The Toyota Product Development System: integrating people, process, and technology*. Productivity Press, Florence, KY.

Morris, L. (1995) 'Creating and maintaining trust', *Training & Development*, 49 (12), 52–4.

Mulder, I. (2004) *Understanding Designers Designing for Understanding*, PhD thesis, Telematica Institute, University of Enschede, The Netherlands.

Osborn, A. F. (1957) *Applied Imagination*, Scribner's, New York.

Othman, A. A. E. (2009) 'Corporate social responsibility of architectural design firms towards a sustainable built environment in South Africa', *Architectural Engineering and Design Management*, 5 (1/2), 36–45 (special issue, 'Design Management for Sustainability', edited by S. Emmitt).

Otter, A. F. H. J. (2005) *Design Team Communication Using a Project Website*, PhD thesis, Bouwstenen 98, Technische Universiteit Eindhoven.

Otter, A. den and Emmitt, S. (2007) 'Exploring effectiveness of team communication: balancing synchronous and asynchronous communication in design teams', *Engineering, Construction and Architectural Management*, 14 (5), 408–19.

Perry, R. B. (1914) 'The definition of value', *Journal of Philosophy and Scientific Methods*, 11 (6), 141–62.

Phillips Report (1950) *Report of a Working Party to the Minister of Works*, HMSO, London.

Poole, M. S. and Hirokawa, R. Y. (1996) 'Communication and group decision-making', in R. Y. Hirokawa and M. S. Poole (Eds), *Communication and Group Decision Making*, Sage, London, pp. 3–18.

Porter, M. E. (1985) *Competitive Advantage: creating and sustaining superior performance*, The Free Press, New York.

Potter, N. (1989) *What is a Designer?: things, places, messages* (third edition), Hyphen Press, London.

Potter, J. and Wetherell, M. (1987) *Discourse and Social Psychology: beyond attitudes and behaviour*, Sage, London.

Renn, O., Blattel-Mink, B. and Kastenholtz, H. (1997) 'Discursive methods in decision making', *Business Strategy and Environment*, 6, 218–31.

Revans, R. W. (1998) *ABC of Action Learning*, Lemos and Crane, London.

Rim, Y. (1966) 'Machiavellianism and decisions involving risk', *British Journal of Social and Clinical Psychology*, 5, 30–6.

Rogers, E. M. (1986) *Communication Technology: the new media in society*, Free Press, New York.

Rogers, E. M. (2003) *Diffusion of Innovations* (fifth edition), Free Press, New York.

Rogers, E. M. and Kincaid, D. L. (1981) *Communication Networks: toward a new paradigm for research*, The Free Press, New York.

Saint, A. (1983) *The Image of the Architect*, Yale University Press, New Haven, CT.

SBi (2005) *SBi Journal* No. 421–042.

Schön, D. A. (1983) *The Reflective Practitioner*, Basic Books, New York.

Schön, D. A. (1985) *The Design Studio*, RIBA Publications, London.

Schön, D. A. (1987) *Educating the Reflective Practitioner*, Jossey-Bass, San Francisco.

Schwartz, S. H. (1994) 'A theory of cultural values and some implications for work', *Applied Psychology*, 48 (1), 23–47.

Senge, P. M. (1990) *The Fifth Discipline: the art and practice of the learning organization*, Doubleday, New York.

Serva, M. A., Fuller, M. A. and Mayer, R. C. (2005) 'The reciprocal nature of trust: a longitudinal study of interacting teams', *Journal of Organisational Behaviour*, 26, 625–48.

Shadish, W. R. (1981) 'Theoretical observations on applied behavioural science', *Journal of Applied Behavioural Science*, 17 (1), 98–112.

Shannon, C. E. and Weaver, W. (1949) *The Mathematical Theory of Communication*, University of Illinois, Urbana.

Simon, E. D. (1944) *The Placing and Management of Building Contracts*, HMSO, London.

Spector, M. D. and Jones, G. E. (2004) 'Trust in the workplace: factors affecting trust formation between team members', *Journal of Social Psychology*, **144** (3), 311–21.

Sperber, D. and Wilson, D. (1986) *Relevance Communication and Cognition*, Blackwell, Oxford.

Stretton, A. (1981) 'Construction communications and individual perceptions', *Chartered Builder*, Summer, 51–3.

Stroop, J. R. (1932) 'Is the judgement of the group better than that of the average member of the group?', *Journal of Experimental Psychology*, **15**, 550–62.

Tajfel, H. and Fraser, C. (Eds) (1978) *Introducing Social Psychology*, Penguin, Harmondsworth.

Taylor, D. W., Berry, P. C. and Block, C. H. (1958) 'Does group participation when using brainstorming facilitate or inhibit creative thinking', *Administrative Science Quarterly*, **3**, 23–45.

Thelen, H. A. (1949) 'Group dynamics in instruction: principle of least group size', *School Review*, **57**, 139–48.

Thyssen, M. H., Emmitt, S., Bonke, S. and Christoffersen, A. K. (2010) 'Facilitating client value creation in the conceptual design phase of construction projects – a workshop approach', *Architectural Engineering and Design Management*, **6** (1), 18–30.

Tombesi, P., Dave, B., Gardiner, B. and Scriver, P. (2007) 'Rules of engagement: testing the attributes of distant outsourcing marriages', *Architectural Engineering and Design Management*, **3** (1), 49–64 (special issue, 'Aspects of Building Design Management', edited by S. Emmitt).

Trenholm, S. and Jensen, A. (1995) *Interpersonal Communication* (third edition), Wadsworth Publishing, London.

Trujillo, N. (1986) 'Implications of interpretive approaches for organization communication research and practice', in L. Thayer (Ed.) *Organization–Communication: emerging perspectives*, Vol. 2: *Communication in Organizations*, Ablex, Norwood, NJ, pp. 46–63.

Tuckman, B. W. (1965) 'Development sequences in small groups', *Psychological Bulletin*, **63**, 384–99.

Volkema, R. J. and Niederman, F. (1995) 'Organizational meetings: formats and information requirements', *Small Group Research*, **26**, 3–24.

Wadleigh, P. M. (1997) 'Contextualizing communication avoidance research', in J. A. Daly, J. C. McCroskey, T. Hopf and D. Ayes (Eds), *Avoiding Communication* (second edition), Hampton Press, Cresskill, NJ, pp. 3–20.

Walker, A. (2007) *Project Management in Construction* (fifth edition), Blackwell Publishing, Oxford.

Wallace, W. A. (1987) *The Influence of Design Team Communication Content upon the Architectural Decision Making Process in the Pre-contract Design Stages*, PhD thesis, Department of Building, Heriot-Watt University.

Waring, T. and Wainwright, D. (2000) 'Interpreting integration with respect to information systems in organizations – image, theory and reality', *Journal of Information Technology*, **15** (2), 131–48.

Wilkinson, P. (2005) *Construction Collaboration Technologies: the extranet evolution*, Taylor & Francis, London.

Womack, J. and Jones, D. (1996) *Lean Thinking: banish waste and create wealth in your corporation*, Simon & Schuster, New York.

Womack, J. P., Jones, D. T. and Roos, D. (1991) *The Machine That Changed the World: the story of lean production*, Harper Business, New York.

Wren, D. A. (1967) 'Interface and interorganizational coordination', *Academy of Management Journal*, **10** (1), 69–81.

Ysseldyke, J. E., Algozzine, B. and Mitchell, J. (1982) 'Special education team decision-making: an analysis of current practice', *Personnel and Guidance Journal*, **60** (5), 308–13.

Yoshida, R. K., Fentond, K. and Maxwell, J. (1978) 'Group decision making in the planning team process: myth or reality?', *Journal of School Psychology*, **16**, 237–44.

Zahrly, J. and Tosi, H. (1989) 'The differential effects of organizational induction process on early work role adjustment', *Journal of Organizational Behavior*, **10**, 59–74.

Index

action learning 156
action research 157
added value 57
agreement 63
application of workshop model 170
architectural fashion 101
assembly of the TPO 109–24;
    procurement choices 112; strategic
    approach 111
asynchronous communication 32
attitude 56, 114

brainstorming 85
bribery 52
briefing, strategic 96
building information modeling (BIM)
    16
burnout 137, 138

category-based trust 49
client: context 93; expertise 94;
    interaction 96; representatives 131;
    types 95
closure of project 139
collaboration: definition 15
collaborative: technologies 16; working
    15
collective communication framework 42
combined meetings and workshops 74
communication 14, 27–43:
    asynchronous 32; channels 33,
    34; closed 35; definitions 28, 29;
    effectiveness 27; formal 34, 35;
    golden rules 43; graphical 39;
    informal 34, 35; interdisciplinary
    31; levels 32; networks 34; open
    35; oral 39; physical 40; process
    29; statutory 35; synchronous 31;
    written 39

complexity 101
conflict 14, 132; life cycle 133;
    management 134; resolution 135;
    types of 133
consensus 14
context 22, 91–107, 162; client
    93; Danish, 162; evolving 92;
    leadership 101; physical 97; process
    management 104; response to 93;
    social 99
contract choices 114
corporate social responsibility (CSR) 53
corruption 51
creative clusters 86
cultural differences 101

decisions 77–89; communication 83;
    extreme views and risk taking 84;
    outside meetings 71
decision making 79; constraints 79;
    group 81; information for 89; norms
    84; stakeholder involvement 80
defensiveness 63
design changes 88
development of the TPO 125–41
disagreement 63, 135
disciplinary language 36
discipline: definition 7
discussion of the workshop model 172
discussions 61–75
dysfunctional meetings 72

electric theatre 40
electronic conferencing 42
ethics 51, 53
evidence-based learning 154
external meetings and workshops 65
extreme views 84

facilitated workshops *see* workshops
fashion: architectural 101; management
    91, 101
fraud 52

gender 50
government-led reports 10
graphical communication 39
group: communication 33; decision
    making 81; development 127;
    goals 129; inconsistent membership
    130; interaction 21; norms 128;
    productivity 129; size 19, 173
groups: definition 18; within TPOs 16
groupthink 87

help seeking 64
hidden agendas 23

implementation 161–75
independent working 21
individual values 55
information and decision making 89
information communication
    technologies (ICTs) 35, 40–3
initial trust 49
integrated: activities, 20; teams 19;
    TPOs 19
integration: definition 19
interaction 164, 166
interactive media 69, 74
interdisciplinary: definition 8, 9;
    groups and teams 16; project
    communication 31
interface management 23, 24
interfaces 7–26; communication
    25; definition 25; organisational
    24; personal 25; recognition 24;
    responsibilities 25
internal meetings and workshops 65
international projects 37, 121
interpersonal communication 32
interpersonal underworld 23
intrapersonal communication 32

language 36
leadership 102, 169; context 101
lean philosophy 57, 161, 163
learning 143–60; action learning 156;
    action research 157; evaluation 153;
    evidence based 154; integrating
    research and practice 147; lifelong
    145; from peers 147; from products

152; from projects 149; reflection
    148; storytelling 158; styles 146
legislative framework 100

managerial fashion 101
managing value 58
mass communication 33
media 38
meetings 64, 66; chairing 68; function
    of 67; making them work 68;
    managing 69; time and place 66
multigroup communication 33

negotiations 62

occupational stress 137
oral communication 39
organizational values 55
outsourced relationships 119

participation: definition 13; within
    TPOs 164
partnering 10, 167
performance measurement 144
philosophy 162, 163
physical context 97
physical representation 40
physicality 97, 98
political factors 101
post-occupancy evaluation (POE) 153
process facilitation 169
process management 104, 105, 164
procurement choices 112
problem solving 78
project: closure 139; cultures 120, 121,
    122; values 55
project manager: behaviour 103, 104;
    changes 131; responsibilities 2;
    selection 118
project websites 41

quality 109
question asking 64
questioning 78

reflection (case study) 173
reflection in action 148
reflective diary 148
relational forms of working 10
relationship building 119
reluctant communicators 30
resourcing of organisations and projects
    119

reviews: post-project 151; through
  project 150
risk management 58
risk taking 84
role-based trust 49

selection: of managers 118; of
  participants 115
selection criteria 116
site selection 99
social and cultural background 22
social context 99
social interaction 98
socio-economic issues 100
stakeholder(s): definition 12
storytelling 158
stress *see* occupational stress
surface behaviour 23
swift trust 51
synchronous communication 31

tasks and procedures 23
teams 16, 17; co-located 17;
  geographical location 18; global 18;
  virtual 18

teamwork 15
temporary project organisations (TPOs):
  assembly 109; culture 121, 122;
  definition 11; development 125;
  integrated 19; trust within 45, 50
trust 45–59, 114; definitions 47; initial
  trust 49; limits and boundaries 47;
  role based 49
trusting stance 49

value 56, 161; added 57; design and
  delivery 165
value management 58
values: based process model 55,
  161,164, 165; definitions 54;
  identification of 54; levels 55;
  personal 54

workshops 64, 166; facilitated 72;
  facilitation of 74, 169; function of
  73; making them work 73; sequence
  167; time and place 66
written communication 39